9/01

MAKING OF THE EARTH

GEOLOGIC FORCES THAT SHAPE OUR PLANET

MAKING OF THE EARTH

GEOLOGIC FORCES THAT SHAPE OUR PLANET

JON ERICKSON

FOREWORD BY DONALD R. COATES, PH.D.

☑ Facts On File, Inc.

MAKING OF THE EARTH
Geologic Forces That Shape Our Planet

Facts On File, Inc.
11 Penn Plaza
New York NY 10001

Library of Congress Cataloging-in-Publication Data

Erickson, Jon, 1948–
   Making of the earth : geologic forces that shape our planet / by Jon Erickson
      p. cm. — (The living earth)
   Includes bibliographical references and index.
   ISBN 0-8160-4276-4
      1. Geodynamics. I. Title

   QE501 .E65 2000
   551.41—dc21   00-03943

Facts On File books are available at special discounts when purchased in bulk quantities for businesses, associations, institutions or sales promotions. Please contact our Special Sales Department at 212/967-8800 or 800/322-8755.

You can find Facts On File on the World Wide Web at **http://www.factsonfile.com**

Text design by Cathy Rincon
Cover design by Nora Wertz
Illustrations by Jeremy Eagle and Dale Dryer, © Facts On File

Printed in the United States of America

VB Hermitage 10 9 8 7 6 5 4 3 2 1

This book is printed on acid-free paper.

CONTENTS

CONTENTS

TABLES

ACKNOWLEDGMENTS

The author thanks the following organizations for providing photographs for this book: the National Aeronautics and Space Administration (NASA), the National Oceanic and Atmospheric Administration (NOAA), the National Park Service, the U.S. Air Force, the U.S. Army Corps of Engineers, the U.S. Coast Guard, the U.S. Department of Agriculture (USDA), the USDA Forest Service, the USDA–Soil Conservation Service, the U.S. Geological Survey (USGS), and the U.S. Navy.

Special thanks also go to Mr. Frank Darmstadt, senior editor at Facts On File, and Ms. Cynthia Yazbek, associate editor, for their contributions in the creation of this book.

FOREWORD

MAKING OF THE EARTH

Among the nine planets of the solar system, Earth is the most unusual. The name *Earth* is, in fact, a misnomer because the abundance of water has provided the unique qualities of our planet. Water covers 70 percent of the global surface, and an additional 10 percent of the land is masked by glaciers. The hydrologic cycle profoundly influences all surface features of the planet. *Making of the Earth* by Jon Erickson places the importance of such phenomena into perspective.

The topographic complexity of Earth's features and processes form the focus of the book. Each of the 10 chapters addresses a major theme fundamental to an understanding of the building blocks that comprise the planet. Chapters 1, 2, 3, and 4 deal with forces within the Earth that create the major architectural patterns of the continents and ocean basins. These internal stresses are of such magnitude that they constitute and form the major framework of the landscape. Careful attention is devoted to global tectonics, the major paradigm that has completely revolutionized geologic thinking of the Earth's structural history. The author provides an up-to-date treatment of how mountains are formed, the forces involved with volcanic and earthquake activity, and their associated processes of faulting, folding, and landform creation.
Chapters 5, 6, 7, 8, and 9 are devoted to dynamic forces that occur on the land surface, such as rivers, oceans, wind, and glaciers. How these processes operate, and the topographic features they create, receive major attention. Indeed,

Making of the Earth vividly chronicles the eternal struggle that the planet exhibits. This history is one of constant subsurface agitation of rock deformation and its subsequent mobilization by tectonic and volcanic pressures. However, once such rocks and features reach the land surface, they are continually attacked by surface processes that mold and change them into an astounding variety of different landforms and terrains. Chapter 10 describes an unusual set of landforms that are different and somewhat bizarre. They are a final reminder of the beautiful array of features that provides Earth with its indelible landscape flavor and personality.

The attractiveness of Mr. Erickson's text is enhanced by several additions. For greater clarity, the illustrations—both diagrams and photographs—provide vivid portrayals that augment word descriptions. The timely bibliography can help the reader if additional information is desired. The glossary is a useful tool for understanding the vocabulary of geology, and the tables give significant figures to quantify important topics. I found the book to be highly readable and an accurate source of historical and current data for anyone interested in the geologic wonders of our planet.

—Donald Coates, Ph.D.

INTRODUCTION

The science of geomorphology is a branch of geology dealing with the study of landforms, the surface features of the Earth. This book discusses the geologic processes that constantly change the face of the planet. Landforms are the basic structures that make up the landscape. They include mountains, volcanoes, plateaus, cliffs, hills, canyons, valleys, basins, and many other geologic formations. All landforms result from a combination of processes that continually build up the land surface and ultimately tear it down.

Landforms are easily recognizable and often express the type of rock that comprises them. Some geologic forms and structures are associated with a particular rock type and can often be identified some distance away. The size, shape, and composition of landforms depend on the nature of the rocks themselves. The knowledge of landforms and structures is therefore fundamentally important for understanding how geology works and aids in interpreting the landscape and geologic history of a region.

Our planet is constantly evolving. The Earth's surface has been fashioned by a variety of geologic forces, offering a wide assortment of structures. Complex activities, such as flowing water and moving waves, rearrange the surface features of the planet. The land surface is sculptured by such formidable geologic processes as plate tectonics, uplift and erosion, powerful earth-moving processes, and ground collapse. Weathering, downslope movement, and river flow work together to continually reshape the continents.

Geologic structures resulting from the interaction of movable tectonic plates, igneous activity, and ground motions provide many unusual land features. Spectacular landscapes are carved by unceasing weathering agents that cut down the tallest mountains and gouge out the deepest ravines. The presence of these geologic phenomena is a tribute to the powerful tectonic forces responsible for creating a large variety of landforms.

This book is meant to introduce to young readers and adults alike this fascinating field of geology. The text covers the following topics: the formation of continents, the evolution of landforms, mountain building, the contribution of volcanoes, the importance of rivers, the changing coastal regions, the development of deserts features, the shaping of arctic geology, landscapes sculpted by glaciers, and unique landforms produced by a variety of geologic forces—including stone monuments, arches, natural bridges, sinkholes, and some of the strangest volcanoes ever encountered.

These topics will provide a basic understanding of one of geology's most fundamental fields. Readers will enjoy this clear and easily readable text that is well illustrated with evocative photographs, drawings, and helpful tables. A comprehensive glossary is provided to define difficult terms, and a bibliography lists references for further reading. Students and geology enthusiasts will gain the tools necessary for discovering an exciting world around them.

1

THE BIRTH OF CONTINENTS

THE FORMATION OF CONTINENTAL CRUST

Continents are unique to the Earth. This chapter will explain how they evolved down through the ages, beginning with the formation of the first continental crust. The continents are constructed out of an assortment of shields, cratons, and terranes. Several times in our planet's history, they combined into supercontinents comprising all the lands of the Earth. The supercontinents, in turn, rifted apart into new continents and oceans, thereby repeating the cycle, until today we have our own unique set of continents.

The Earth has undergone many dramatic changes over geologic time (Table 1). The continental crust originated as early as 4 billion years ago during Precambrian time when slices of rock combined to form the nuclei of the continents, upon which all other rocks were deposited. During the Paleozoic era, between about 360 and 270 million years ago, the continents assembled into the supercontinent Pangaea (Fig. 1), which comes from Greek and means "all lands." Surrounding this huge landmass was a large body of water called

TABLE 1   THE GEOLOGIC TIME SCALE

Era	Period	Epoch	Age (millions of years)	First life forms	Geology
		Holocene	0.01		
	Quaternary				
		Pleistocene	3	Humans	Ice age
Cenozoic		Pliocene	11	Mastodons	Cascades
		Neogene			
		Miocene	26	Saber-toothed tigers	Alps
	Tertiary	Oligocene	37		
		Paleogene			
		Eocene	54	Whales	
		Paleocene	65	Horses, Alligators	Rockies
	Cretaceous		135		
				Birds	Sierra Nevada
Mesozoic	Jurassic		210	Mammals	Atlantic
				Dinosaurs	
	Triassic		250		
	Permian		280	Reptiles	Appalachians
	Pennsylvanian		310		Ice age
				Trees	
	Carboniferous				
Paleozoic	Mississippian		345	Amphibians	Pangaea
				Insects	
	Devonian		400	Sharks	
	Silurian		435	Land plants	Laursia
	Ordovician		500	Fish	
	Cambrian		570	Sea plants	Gondwana
				Shelled animals	
			700	Invertebrates	
Proterozoic			2500	Metazoans	
			3500	Earliest life	
Archean			4000		Oldest rocks
			4600		Meteorites

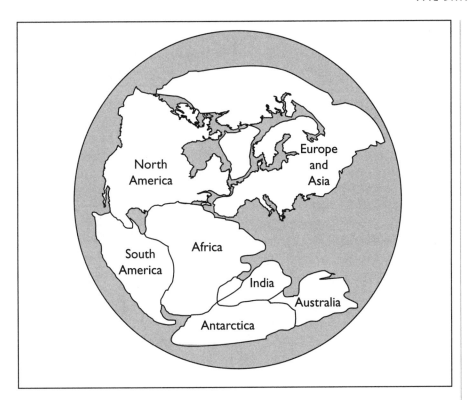

Figure 1 *The supercontinent Pangaea 280 million years ago.*

Panthalassa, from the Greek meaning "universal sea." This superocean stretched practically uninterrupted over the entire planet while all the continents huddled to one side.

Pangaea rifted apart, forming the present continents and oceans about 180 million years ago during the Mesozoic era. Extensive volcanism followed the continental breakup and provided an extraordinary warm climate during the reign of the dinosaurs. The Cenozoic era, from 65 million years ago to the present, witnessed an extraordinary growth of mountain ranges as the continents drifted toward their present locations. Most of the continental mass gathered around the North Pole, prompting a colder climate and a series of ice ages. Perhaps in another 200 million years, the entire continental crust will reunite in a new supercontinent, and the processes of rifting and patching will continue far into the future.

CONTINENTAL CRUST

The Earth's crust formed in a burst of creation, starting as early as 4 billion years ago. During the first half-billion years, the Earth was in a fiery turmoil;

Figure 2 *The heavily cratered Marius Hills region on the Moon.*

(Photo by D. H. Scott, courtesy of USGS)

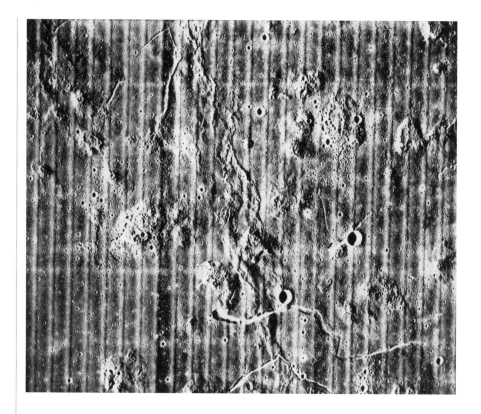

rocks solidifying on the surface soon remelted from the intense heat. Impact friction from a massive bombardment of meteorites between 4.2 and 3.8 billion years ago also melted portions of the crust. Thousands of asteroids some 50 miles wide or more that were left over from the creation of the solar system pounded the formative planet. The impact basins created by the asteroids' collisions covered half the crust, giving it an appearance resembling the heavily cratered regions of the moon (Fig. 2), where the meteorite impacts have been much better preserved than those on Earth.

A permanent crust did not develop until around 4 billion years ago. The original crust consisted of lava flows that erupted on the surface long before the ocean basins filled with water. Embedded in the thin basaltic crust were massive granitic blocks called rockbergs. These rocks assembled into microcontinents less dense than the underlying basalt, which buoyed them to the surface. The bulk of the crust composed of oxygen, silica, and aluminum comprised the granitic and metamorphic rocks in the cores of the continents.

The existence of ancient metamorphosed granites in northwest Canada dating 4 billion years old called Acasta Gneiss indicates that substantial continental crust comprising as much as 10 percent of the present landmass was present by this time. The metamorphosed marine sediments of the Isua

Formation in a remote mountainous region in southwest Greenland suggest the presence of a saltwater ocean by at least 3.8 billion years ago. Older rocks are found in Antarctica and Africa, but few date beyond 3.7 billion years. This implies that much of the early crust was recycled into the mantle. Therefore, only a fraction of today's continental mass had formed by this time.

No significant landmasses had yet formed, and only thin slices of crust wandered across the watery face of the Earth. They were driven by rapid convective or circular motions in the mantle, with a heat flow three times greater than that today. The continued loss of internal heat slowed the convective flow, enabling the lighter rock material to migrate toward the surface, where it formed a refuse similar to the slag on molten iron ore. In essence, the crust was made from the waste products of the mantle.

The reworking of this primitive crust as it descended into the mantle and remelted during tectonic activity produced the first granites. Rocks thrust deep into the hot mantle both metamorphosed and either changed their crystalline structure or melted entirely. They then became molten rock known as magma. The buoyant magma rose to the surface in giant blobs called diapirs, which comes from the Greek word *diapeirein* and means "to pierce." Some magma erupted onto the surface, spewing from a profusion of volcanoes. The rest remained buried in the crust, where the magmatic intrusion created large granitic bodies called plutons (Fig. 3).

The early continents were highly mobile crustal fragments, constantly colliding with one another. Most continental crust developed when individ-

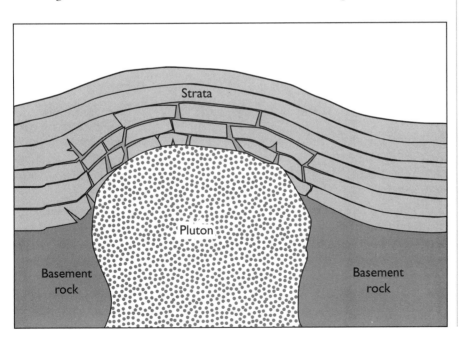

Figure 3 *A pluton is produced by the intrusion of a magma body into overlying rocks.*

ual crustal plates collided, deforming the crust over a broad area. As the Earth's interior continued to cool, the continents slowed their erratic wanderings and began to combine into a dozen or more protocontinents or precontinental masses.

Several times in Earth's history, the protocontinents came together into larger supercontinents. The events that formed global continents occurred roughly 2.9 to 2.6 billion years ago, 1.9 to 1.7 billion years ago, 1.3 to 1.1 billion years ago, and 0.5 to 0.3 billion years ago. Extensive volcanic activity and the formation of plutons within the crust built up the continental interiors, while erosion and sedimentation expanded the continental margins outward.

The crust (Table 2) comprises less than 1 percent of the Earth's radius and about 0.3 percent of its mass. It is composed of ancient continental rocks and comparatively young oceanic rocks. The continental crust resembles a layer cake with sedimentary rocks on top, granitic and metamorphic rocks in the middle, and basaltic rocks on the bottom. This structure is like a jelly sandwich, with a pliable middle layer placed between a solid upper crust and a hard lithosphere, the rigid uppermost layer of the mantle. Most of the continental rocks originated from volcanoes stretching across the ocean that were drawn together by plate tectonics, which comes from the Greek *tekton* and means "builder."

TABLE 2   CLASSIFICATION OF THE EARTH'S CRUST

Environment	Crust type	Tectonic character	Thickness in miles	Geologic features
Continental crust overlying stable mantle	Shield	Very stable	22	Little or no sediment, exposed Precambrian rocks
	Midcontinent	Stable	24	
	Basin and range	Very unstable	20	Recent normal faulting, volcanism, and intrusion; high mean elevation
Continental crust overlying unstable mantle	Alpine	Very unstable	34	Rapid recent uplift, relatively recent intrusion; high mean elevation
	Island arc	Very unstable	20	High volcanism, intense folding and faulting
Oceanic crust overlying stable mantle	Ocean basin	Very stable	7	Very thin sediments overlying basalts, no thick Palaeozoic sediments
Oceanic crust overlying unstable mantle	Ocean ridge	Unstable	6	Active basaltic volcanism, little or no sediment

When including continental margins and small shallow regions in the ocean, the continental crust covers about 45 percent of the Earth's surface. It varies from 6 to 45 miles thick and rises on average about 4,000 feet above sea level. The thinnest parts of the continental crust lie below sea level on continental margins, and the thickest portions underlie mountain ranges.

Today, the Earth's crust is relatively thin compared with that of the moon and the other inner, rocky planets. A thick, buoyant crust could not exist on this planet because it would have remelted from heat generated by a high concentration of radioactive elements as well as the high temperatures produced by the weight of the overlying rocks. A thick crust would insulate the mantle, holding in its heat and raising temperatures to the melting point of surface rocks. A thick crust also would be highly unstable, resulting in a massive overturn and reabsorption into the mantle. The lack of any evidence of the first 700 million years of Earth's history, known as the Hadean or Azoic era, is a plausible testimony to such an occurrence. Once the seeds of the continental crust were firmly planted, however, the building of the continental shields soon followed.

CONTINENTAL SHIELDS

Basement rocks underneath the surface known as granulites form the nuclei around which the continents formed. They are exposed in broad, domelike structures called shields (Fig. 4). These are extensive, uplifted areas essentially bare of recent sedimentary deposits and contain only thin soils. Surrounding

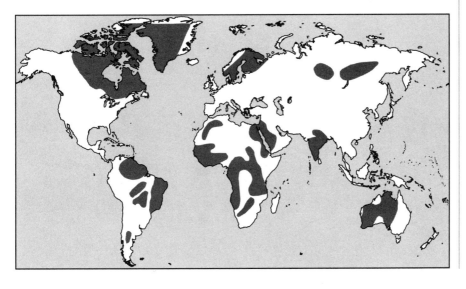

Figure 4 The Precambrian shields comprise the oldest rocks on Earth.

the shields are broad, shallow depressions of basement rock covered with near-ly flat-lying sedimentary strata called continental platforms.

The best-known of the dozen or so continental shields, which contain the oldest rocks on Earth, are the Canadian Shield in North America and the Fennoscandian Shield in Europe. These regions are fully exposed where flowing ice sheets eroded their sedimentary cover during the Pleistocene Ice Age. More than a third of Australia is Precambrian shield. Sizable shields lie in the interiors of Africa, South America, and Asia as well.

Early in the Precambrian, fragments of land—the nuclei upon which the continental landmasses formed—were separated by marine basins, where lava and volcanic sediments accumulated. The deposits metamorphosed into greenstone belts dispersed throughout the shields. Greenstones are a mixture of metamorphosed lava flows and sediments derived from volcanic island arcs on the edges of subduction zones, where the oceanic crust plunges into the mantle. These lava flows and sediments were caught between colliding continents, forming greenstones (Fig. 5).

Figure 5 *An island arc, created by a subduction zone where two plates converge, is caught between colliding continents*

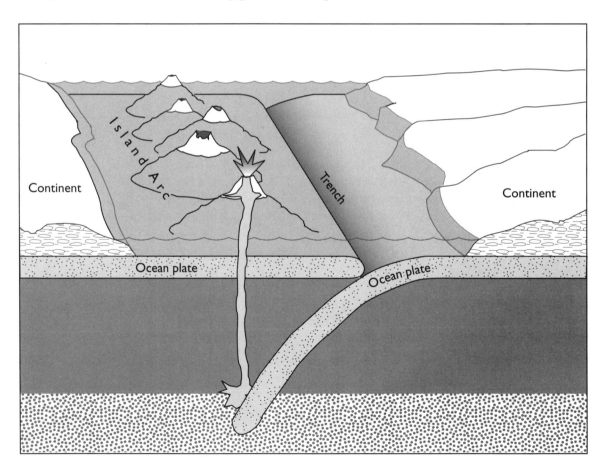

Greenstones have no modern equivalents, therefore the geologic conditions under which they formed differed greatly from those observed today. The Earth's crust was warmer and more pliable, and crustal plates were more mobile. Active tectonic forces in the mantle broke open the crust and injected magma deep into the crustal plates. These large-scale magmatic intrusions along with massive meteorite bombardments characterized the unusual geology of the time.

Ophiolites, from the Greek word *ophis* meaning "serpent," are scattered throughout the greenstone belts. They are slices of ocean floor shoved onto the continents by drifting plates. Ophiolites range in age up to 3.6 billion years old. Pillow lavas (Fig. 6), tubular bodies of basalt erupted on the ocean floor, also appear in the greenstone belts. These deposits are among the best evidence for plate tectonics operating early in the Precambrian. Therefore, continents have been on the go practically from the very beginning.

During plate collisions, ophiolites formed when the oceanic crust peeled off and was plastered against the continents. Blueschists consist of metamorphosed rocks of subducted oceanic crust, which has been forced into the man-

Figure 6 *Pillow lava exposed on the south side of Wadi Jizi, northeast of Suhaylah, Sultanate of Oman.*

(Photo by E. H. Bailey, courtesy of USGS)

tle at subduction zones on the ocean floor. They formed under high pressure. These blueschists were also thrust onto the continents, providing further evidence for early tectonic activity. The continental collisions resulted in linear formations of greenish volcanic rocks along with light-colored masses of granite and gneiss, common igneous and metamorphic rocks comprising the bulk of the continents (Fig. 7). Many ophiolites also contain rich ores that provide important mineral resources the world over, such as the Apennines of northern Italy, the Urals of Russia, and the Andes of South America.

The greenstone belts span an area of several hundred square miles and are surrounded by immense expanses of gneiss, the metamorphic equivalent of granite and the predominant Archean rock type formed between 4 and 2.5 billion years ago. Their color derives from chlorite, a greenish micalike mineral. The best-known greenstone belt is the Swaziland sequence in the Barberton Mountain Land of southeastern Africa. It is nearly 12 miles thick and more than 3 billion years old.

Figure 7 A granite pegmatite dike in Precambrian gneiss in Bear Creek Canyon, Jefferson County, Colorado.

(Photo by J. R. Stacy, courtesy of USGS)

Geologists are particularly interested in greenstone belts because they contain most of the world's gold. India's Kolar greenstone belt holds the richest gold deposits. It is some 3 miles wide and 50 miles long, and it formed when two plates clashed about 2.5 billion years ago. In Africa, the best deposits are in rocks as old as 3.4 billion years. Most South African gold mines are found in greenstone belts. In North America, the best gold mines are in the Great Slave region of northwest Canada, where well over a thousand deposits are known.

STABLE CRATONS

Together, the shields and their surrounding basement rock, called platforms, comprise the stable cratons, which lie in the continental interiors. Cratons consist of ancient igneous and metamorphic rocks, whose composition is remarkably similar to their modern equivalents. The existence of cratons early in Earth's history suggests a fully operating rock cycle was already in place.

Continents were set adrift since they made their first appearance soon after the Earth's formation. The development of continental crust was well underway by this time, as manifested by a 4-billion-year-old metamorphosed granite known as Acasta Gneiss in the Great Slave region of Canada's Northwest Territories. Its discovery leaves little doubt that at least small patches of continental crust were present during the first billion years of the Earth's existence.

Only three sites, located in Canada, Australia, and Africa, contain rocks exposed on the surface during the early Precambrian that have not changed significantly throughout geologic time. Most other rocks have eroded, were metamorphosed, or melted entirely, leaving only a few untouched. The best exposure of Precambrian metamorphic rocks in the United States is the 1.8-billion-year-old Vishnu Schist on the floor of the Grand Canyon (Fig. 8).

Slivers of granitic crust combined into stable bodies of basement rock, forming the cratons, the nuclei upon which all other rocks were deposited. The cratons were the first pieces of continental crust to appear. The original cratons arose within the first 1.5 billion years and totaled only a fraction of today's landmass. The rocks originated from the intrusion of magma into the primitive oceanic crust. They consisted of highly altered granite along with metamorphosed marine sediments and lava flows.

The cratons numbered in the dozens and ranged in size from larger than India to smaller than Madagascar, which are themselves continental fragments. The cratons moved about freely on the molten rocks of the upper mantle. This mantle layer is called the asthenosphere, which comes from the Greek *asthenes*

Figure 8 *The Precambrian Vishnu Schist on the bottom of Grand Canyon, Arizona. 1, Vishnu Schist; 2, Grand Canyon Series; 3, Tonto Group.*

(Photo by E. D. McKee, courtesy of USGS)

and means "weak." The cratons were independent minicontinents whose collisions crumpled the leading edges, forming short parallel mountain ranges.

As the mantle gradually cooled, the cratons slowed their erratic wanderings and combined into larger landmasses, continuously building up the continental crust. Volcanoes were prevalent on the cratons, and lava and ash added to the continental masses. Magmatic intrusions of molten crustal rocks that were recycled through the upper mantle supplied new material to the interior of the cratons. The continents grew rapidly, peaking about 2.5 billion years ago. At that time, they comprised about 80 percent of the present continental landmass and occupied up to a quarter of the Earth's surface. Plate tectonics was fully functional, and much of the world as we know it began to take shape.

In the late Precambrian, all cratons coalesced into a supercontinent several thousand miles wide. The collisions forced up mountain ranges. The seams joining the landmasses remain as cores of ancient mountains called orogens, from the Greek *oros* meaning "mountain." The interval was possibly the most energetic period of tectonic activity, which rapidly built new continental crust.

Over the past half-billion years, about a dozen individual continental plates welded together to form Eurasia (Fig. 9). Eurasia is the youngest and largest modern continent still being pieced together; chunks of crust arrive from the south, riding on highly mobile tectonic plates. The African and South American continents aggregated from mobile cratons about 700 million years ago. North America, possibly the oldest continent, assembled from seven cratons around 2 billion years ago (Fig. 10), forming central Canada and north-central United States. Half a billion years ago, North America was a lost continent, drifting on its own, while most other landmasses combined into a supercontinent.

Continental collisions continued adding new crust to the growing proto–North American continent. A large portion of the continental crust underlying the United States from Arizona to the Great Lakes to Alabama formed in a surge of crustal generation unequaled in North America since. The assembled continent was stable and resilient, able to withstand a billion years of jostling and rifting. It continued to grow as bits and pieces of continents and island arcs adhered to its margins.

Large masses of volcanic rock near the eastern edge of the North American continent indicate that around 750 million years ago, the continent sat at the core of a supercontinent called Rodinia (Fig. 11), which is Russian

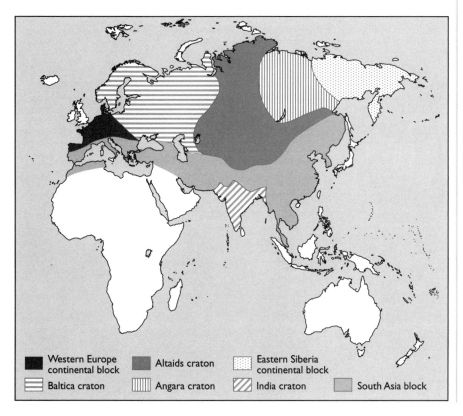

Figure 9 The cratons that comprise Eurasia.

Western Europe continental block

Altaids craton

Eastern Siberia continental block

Baltica craton

Angara craton

India craton

South Asia block

Figure 10 *The cratons that comprise North America.*

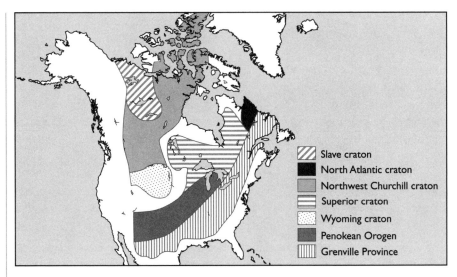

Slave craton
North Atlantic craton
Northwest Churchill craton
Superior craton
Wyoming craton
Penokean Orogen
Grenville Province

Figure 11 *The supercontinent Rodinia 700 million years ago.*

for "motherland." As the supercontinent's interior heated and erupted with volcanism between 630 and 560 million years ago, the warm, weakened crust split apart into possibly four or five major continents whose geographies would be unrecognizable today.

EXOTIC TERRANES

The cratons are a patchwork of crustal blocks that combined into geologic collages known as terranes, most of which are bounded by faults and are distinct from their geologic surroundings. The boundaries between two or more terranes, called suture zones, are commonly marked by ophiolite belts. These consist of ancient oceanic crust shoved onto the continents by drifting plates. The composition of most terranes is similar to that of a volcanic island or undersea plateau. Others comprise a consolidated conglomerate of pebbles, sand, and silt that accumulated in an ocean basin between colliding crustal fragments.

Most terranes are generally elongated bodies that deformed when colliding and accreting to a continent. The assemblage of terranes in China is being stretched and displaced in an east-west direction as India continues to press against southern Asia after colliding with the mainland some 45 million years ago. The buckling crust raised the Himalaya Mountains and the broad Tibetan Plateau, the largest topographic upland on Earth. Asia was elongated to accommodate the northward advancement of India. A belt of ophiolites marks the boundary between the sutured continents.

Terranes take a variety of shapes and sizes ranging from small slices of crust to subcontinents as large as India, itself a single great terrane. They range in age from well over a billion to less than 200 million years old. The ages of the terranes were determined by studying entrained fossil radiolarians (Fig. 12). These are marine protozoans with skeletons made of silica. They were abundant and existed from about 500 million to 160 million years ago. Different species also defined specific regions of the ocean where the terranes originated.

Generally, terranes are fault-bounded blocks with geologic histories separate from those of neighboring terranes and of adjoining continental masses. Many have traveled considerable distances before finally colliding with continental margins. For example, some North American terranes originated in the western Pacific and traversed thousands of miles eastward. In a similar fashion, Eurasia is accumulating pieces of crust arriving from the south. At their usual rate of travel, terranes could circle the entire globe in a mere half-billion years.

Prior to 250 million years ago, during the assembly of the supercontinent Pangaea, the western edge of North America ended near present-day Salt

Lake City. Over the last 200 million years, North America has expanded some 25 percent during a major pulse of crustal growth following the breakup of Pangaea. The entire Pacific Coast from the Baja California Peninsula to the tip of Alaska was grafted onto the continent by the piecemeal addition of crustal blocks.

Much of western North America was assembled from island arcs and other crustal debris skimmed off the 7,500-mile-wide Pacific plate as the North American plate headed westward. Lying in the middle of Wyoming is a nearly complete slice of oceanic crust that was shoved up onto the continent by drifting plates some 2.7 billion years ago. Basaltic seamounts accreting to the margin of Oregon moved from nearby offshore. Similar rock formations consisting of three distinct rock units located around San Francisco, California arrived from halfway across the Pacific Ocean. Northern California is a jumble of crust assembled more than 100 million years ago, comprised of rock formations that originated as far away as 2,500 miles.

Alaska is an agglomeration of terranes, some of which are well exposed in the Brooks Range (Fig. 13), the spine of northern Alaska. This area consists of great sheets of crust stacked one onto another. The entire state is unique because it is an assemblage of some 50 terranes set adrift over the past 160 million years by the wanderings and collisions of crustal plates, fragments of which are still arriving from the south. Some 70 million years ago, Vancouver Island, Canada was nestled along the coast of what is now Baja California. During the next 50 million years, the Pacific plate carrying California west of the San Andreas Fault will slide northward relative to the North American plate at a rate of one or two inches per year. This plate will finally come to rest at the continental margin off Alaska, adding another piece to the puzzle.

To illustrate the distance some terranes can travel, the Alexander Terrane, which makes up a large portion of the Alaskan Panhandle, began as part of eastern Australia some 500 million years ago. Beginning about 375 million

Figure 13 *Steeply dipping Paleozoic rocks of the Brooks Range, Anaktuvuk District, Northern Alaska.*

(Photo by J. C. Reed, courtesy of USGS)

years ago, it broke free of Australia, traversed the Pacific Ocean, stopped briefly at the coast of Peru, and sliced past California, swiping part of the Mother Lode gold belt. It finally collided with the upper North American continent around 100 million years ago.

The accreted terranes played a major role in the formation of mountain chains along convergent continental margins that mark the collision of lithospheric plates. Geologic activity around the Pacific Rim was responsible for practically all mountain ranges facing the Pacific Ocean and the island arcs along its perimeter. Therefore, by all accounts, the Andes Mountains in western South America should not be there. Most of the world's great mountain ranges were created when plate tectonics slammed two continents together. The Andes, however, appear to have been thrust upward by the accretion of oceanic plateaus along the continental margin of South America as the continent overrode a Pacific plate.

When terranes collide and accrete to a continent, the crustal movements modify their shapes. Many terranes in western North America have rotated clockwise as much as 70 degrees or more, with the oldest terranes having the most rotation. Along the mountain ranges in western North America, the terranes are elongated bodies due to the slicing of the crust by a network of faults with a northwest trend. The most active of these faults is California's San Andreas, which has been displaced some 200 miles over the last 25 million years (Fig. 14). The Gulf of California, separating the Baja California Peninsula from mainland Mexico, is a continuation of the San Andreas Fault system. The landscape is literally being torn apart while opening one of the youngest and richest seas on Earth. It began rifting some 6 million years ago, offering a new outlet to the sea for the Colorado River, which thereafter began carving out the Grand Canyon.

CONTINENTAL RIFTS

The continental crust is broken up by faults that often leave long, linear structures when exposed on the surface. The continental crust and underlying lithosphere are generally between 50 and 100 miles thick. Therefore, breaking open a continent would appear to be a monumental task. During the rifting of continents into separate plates, thick lithosphere must somehow give way to thin lithosphere.

The transition from a continental rift to an oceanic rift is accompanied by block faulting. Huge blocks of continental crust drop down along extensional faults, where the crust is diverging. Convection currents rising through the mantle spread out under either side of the lithosphere, pulling the thinning crust apart to form a deep rift. While the rift proceeds across the continent, large earthquakes rattle the region.

Figure 14 *A stream off-set of one-fourth mile on the San Andreas Fault in Carrizo Plains, San Luis Obispo County, California.*

(Photo by R. E. Wallace, courtesy of USGS)

Volcanic eruptions are also prevalent due to the abundance of molten magma rising from the mantle as it nears the surface. Since the crust beneath a rift is only a fraction of its original thickness, magma finds an easy way out. As the crust continues to thin, magma approaches the surface, causing extensive volcanism. A marked increase in volcanic activity during the early stages of many rifts produces vast quantities of lava that flood the landscape.

When a continent fragments, the rift spreads farther apart and floods with seawater, eventually becoming a new sea. As the rift continues to widen and deepen, it is replaced by an oceanic spreading ridge system. Hot material from the mantle wells up through the rift to form new oceanic crust between the two separated segments of continental crust.

The rifting process begins with hot-spot volcanism at rift valleys, which explains the large increase in volcanic activity during the early stages of continental rifting. The hot spots, which are mantle plumes originating from deep

within the mantle, weaken the crust by burning holes through it like geologic blowtorches. The hot spots are connected by rifts, along which the continent eventually breaks up. More than 2 million cubic miles of molten lava erupt when a continent rifts apart over a hot mantle plume, producing enough basalt to cover the entire United States to a depth of half a mile.

Mantle material rises in giant plumes and underplates the crust with basaltic magma, further weakening it and causing huge blocks to drop down to form a series of grabens, which comes from the German meaning "ditch." Grabens are long, trenchlike structures formed by the downdropping of large blocks of crust bounded by normal faults. Some grabens are expressed in the surface topography as linear structural depressions considerably longer than they are wide.

The highland areas flanking the grabens often consist of horsts, which is from German and means "ridge." They are long, ridgelike structures produced when large blocks of crust bounded by reverse faults are upraised with little or no tilting. Horsts and grabens combine to form long parallel mountain ranges and deep valleys such as the Great East African Rift, the Rhine Valley in Germany, the Dead Sea Valley in Israel, the Baikal Rift in southern Russia, and the Rio Grande Rift in the American Southwest (Fig. 15), which slices northward through central New Mexico into Colorado.

The Basin and Range is a unique province of North America (Fig. 16). It is a 600-mile-wide region of fault-block mountain ranges bounded by high-angle normal faults. The province covers southern Oregon, Nevada, western Utah, southeastern California, the southern portions of Arizona and New Mexico, and northern Mexico. The crust has been broken, tilted, and upraised nearly a mile above the basin, radically transforming a land of once gently curving hills into nearly parallel ranges of high mountains and deep basins.

The mountains formed at the tops of the fault blocks, and V-shaped valleys formed at the bases. The valley floors flattened as they filled with sediments washed down from the mountains. The stretching and faulting gradually moved westward, finally peaking in Death Valley, California about 3 million years ago.

The Great Basin of Nevada and Utah is a 300-mile-wide closed depression created by the stretching and thinning of the crust. The crust in the entire

Figure 16 *The Basin and Range Province of the western United States.*

(Photo courtesy of General Electric and NASA)

Basin and Range Province is actively spreading apart by forces originating in the mantle. Most of the deformation results from extension occurring along a line running approximately northwest to southeast. The tectonic changes ushered in volcanism in places where the crust was weakening, allowing magma to well up to the surface.

As the crust continues to separate, some blocks sink, forming grabens separated by horsts. About 20 horst-and-graben structures extend from the 2-mile-high scarp of the Sierra Nevada in California to the Wasatch Front, a major fault system running roughly north-south through Salt Lake City, Utah. The horst-and-graben structures trend northwestward, roughly perpendicular to the movement of the blocks of crust that are spreading apart.

Resting between the ranges are basins, where dry lake bed deposits commonly occur because many low-lying areas once contained lakes. Utah's Great Salt Lake and the Bonneville Salt Flats are good examples. The region is literally stretching apart as the crust is weakened by a series of downdropped blocks. Consequently, 15 million years ago, the present sites of Reno, Nevada and Salt Lake City, Utah were 200 to 300 miles closer together. They are spreading apart about an inch per year.

Death Valley (Fig. 17), at 280 feet below sea level, is the lowest point on the North American continent. The area, at one time, was elevated several thousand feet but collapsed when the continental crust thinned from extensive block faulting in the region. The Great Basin is now only a remnant of a broad belt of mountains and high plateaus that was down-dropped after the crust pulled apart during the growth of the Rocky Mountains. The Andes Mountains could suffer a similar fate as the plate upon which they stand thins out and collapses. Without buoyant support from below, the mountains will cease rising and will be torn down by erosion.

Africa's rift valleys are literally tearing the continent apart. They are a complex system of parallel horsts, grabens, and tilted fault blocks. The border faults have slipped as much as 8,000 feet. The eastern rift zone lies east of Lake Victoria and extends 3,000 miles from Mozambique to the Red Sea. The western rift zone lies west of Lake Victoria and extends 1,000 miles northward. The rift just north of Lake Victoria holds Lake Tanganyika, the second deepest lake in the world. Russia's Lake Baikal, at about 5,700 feet deep and 375 miles long, is the record holder. It fills the Baikal Rift Zone, a crack in the crust similar to the East Africa Rift.

The East African Rift Valley (Fig. 18), marking the boundary between the Nubian plate to the west and the Somalian plate to the east, has not yet fully ruptured. Therefore, it provides the best evidence for the rifting of continents. The region extends from the shores of Mozambique to the Red Sea, where it splits to form the Afar Triangle in Ethiopia. The rift is a complex system of tensional faults, indicating the continent is in the initial stages of rup-

ture. Once the region finally breaks up, the continental rift will be replaced by an oceanic rift.

This transition is presently occurring in the Red Sea, which is rifting from north to south. The Gulf of Aden is a young oceanic rift between the fractured continental blocks of Arabia and Africa. These blocks have been diverging for more than 10 million years. The breakup of North America, Eurasia, and Africa beginning about 170 million years ago was probably initiated by an upwelling of basaltic magma similar to that presently taking place under the Red Sea and the East African Rifts.

Much of the East African Rift system has been uplifted thousands of feet by an expanding mass of molten magma lying just beneath the crust. This heat source is responsible for numerous hot springs and volcanoes along the rift valley. Many of the largest and oldest volcanoes in the world stand nearby,

Figure 17 *Death Valley, California, showing sparse vegetation and roving sand dunes.*

(Courtesy of National Park Service)

Figure 18 *The East African Rift system (dotted lines), where the continent is being pulled apart by plate tectonics.*

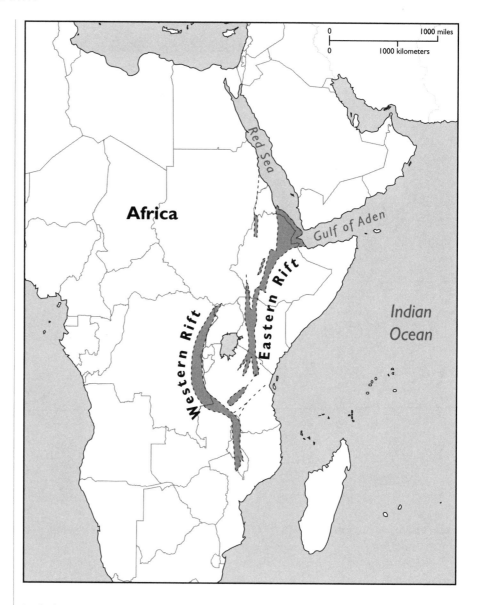

including Mounts Kenya and Kilimanjaro, whose Kibo Peak—at 19,340 feet—is the highest point on the African continent.

The next chapter, explains the geologic processes that shape the continents.

2

TECTONIC PROCESSES
THE SHAPING OF THE PLANET

This chapter examines the powerful forces that shape our planet. These include plate tectonics, volcanism, earthquakes, earth movements, erosion, and sedimentation. We live on a remarkably dynamic planet, with majestic mountains, colossal canyons, violent volcanoes, enormous earthquakes, and other powerful, earth-moving forces. The Earth offers many spectacular landscapes, sculptured by vigorous weathering agents that cut down the tallest mountains and gouge out the deepest ravines. Rivers transport the sediments to the sea, where they accumulate into thick deposits that lithify into solid rock.

These activities are expressions of plate tectonics. Convective motions in the mantle move the continents around the globe, making the Earth a living planet in every respect. A jumble of crustal plates continuously in motion (along with strong erosional forces) provides a myriad of geologic features found nowhere else in our solar system.

GLOBAL TECTONICS

The Earth's outer shell is fashioned into several large, mobile, tectonic plates (Fig. 19), comprising the crust and the upper brittle mantle called the

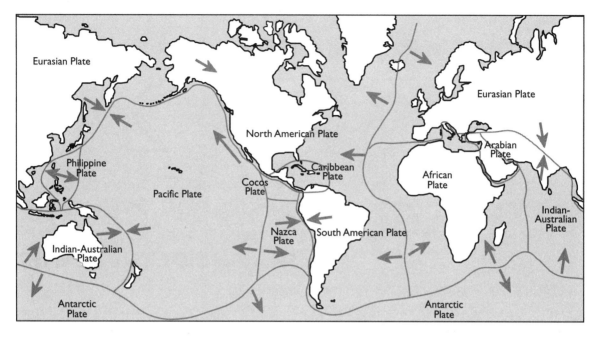

Figure 19 *The lithospheric plates that comprise the Earth's crust.*

lithosphere. The lithospheric plates ride on the semimolten layer of the upper mantle called the asthenosphere, which mechanically behaves as a fluid. The shifting plates range in size from hundreds to tens of millions of square miles and average about 60 miles thick. This unique structure, absent on all the other planets, is responsible for the operation of plate tectonics, as the interactions of lithospheric plates continuously shape the surface of the Earth.

Most plate tectonic activity occurs in the ocean, where new oceanic crust is created at midocean ridges and old oceanic crust is destroyed at subduction zones. The plate boundaries are volcanic rifts, transform faults, and subduction zones. Midocean rifts, where basalt wells up from within the mantle, generate new lithosphere in a continuing process of seafloor spreading. As lithospheric plates diverge, the rift fills with molten magma from the asthenosphere, creating new oceanic crust at spreading ridges.

The ridge system snakes 40,000 miles around the globe, making it the longest uninterrupted structure on Earth. The vast majority of molten magma cools and bonds to the edges of separating plates. This process creates new oceanic crust. Magma also spills out onto the surface of the ridges, producing submarine volcanic eruptions. As the seafloor spreads apart, the plates slide past each other along transform faults, ranging from a few miles to several hundred miles in length.

The subduction zones form deep-sea trenches on the ocean floor, where old oceanic crust sinks into the mantle to produce new basalt in a continuous cycle of crustal regeneration (Fig. 20). The subducting plates also carry the surrounding viscous mantle down with them into the Earth's interior, with the line of subduction marked by the deepest trenches. If tied end to end, the subduction zones would stretch completely around the world. Most subduction zones surround the Pacific Basin, where plate subduction is responsible for the intense seismic activity fringing the Pacific Ocean in a region called the circum-Pacific belt. This zone also coincides with the Ring of Fire, renown for its extensive volcanic activity.

The lithospheric plates transport continental crust around the surface of the Earth like ships frozen in Arctic pack ice. The continents surrounding the Atlantic are driven by deep-mantle flow, whereas those encircling the Pacific are driven by plate subduction. Therefore, the force of gravity is mostly responsible for the drifting of the continents. In this manner, pull at subduction zones is favored over push at spreading ridges to move huge blocks of crust around the Earth.

The bulk of the crust is comprised of granitic and metamorphic rocks buoyed onto the surface of the mantle because of their lower density. When two plates collide, the less buoyant oceanic crust subducts under lighter continental crust or younger oceanic crust, whose higher temperature makes it more buoyant. As the oceanic crust cools, thickens, and descends into the mantle, it remelts and generates new molten magma.

Collisions between plates create volcanic islands in the ocean and raise mountain ranges on the continents. When an oceanic plate subducts under the

Figure 20 The plate tectonics model, in which lithosphere created at midocean ridges is subducted into the mantle at deep-sea trenches, causing the continents to drift around the Earth.

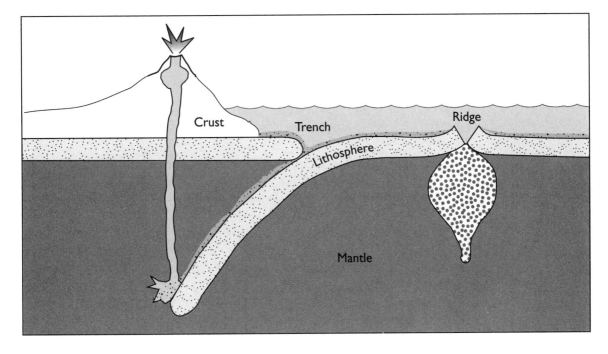

leading edge of a continental plate, it forms sinuous mountain chains—such as the Andes along western South America—and volcanic mountain ranges—such as the Cascades in the Pacific Northwest of North America (Fig. 21). Plate collisions build supercontinents; the breakup of plates creates new continents and oceans.

The plates on which the Americas, Eurasia, and Africa ride spread apart at the Mid-Atlantic Ridge approximately an inch per year or about as fast as a fingernail grows. As the Atlantic Basin widens, the bordering continents separate at the expense of the Pacific Basin. Subduction zones around the Pacific Ocean assimilate old lithosphere, thereby shrinking the Pacific and adjacent plates.

Old oceanic crust, composed of basalts originating at spreading ridges and sediments washed off nearby continents and islands, subducts into the mantle. On the way deep into the Earth's hot interior, the oceanic crust and underlying lithosphere melt. The molten magma rises toward the surface in huge plumes. When the magma reaches the base of the crust, it supplies new molten rock to magma chambers beneath volcanoes and granitic bodies called plutons, from which mountains often form. Magma erupting on the surface

Figure 21 *Mount Shasta, Cascade Range, Siskiyou County, California.*

(Photo by C. D. Miller, courtesy of USGS)

builds volcanic structures, including mountains and broad plateaus on the continents and islands in the sea.

VOLCANOES AS BUILDERS

Most volcanoes occur during movements of crust where lithospheric plates interact with each other. When one plate subducts or descends under another, the lighter rock component melts and rises to feed active volcanoes. Volcanism associated with subduction zones produces island arcs in the ocean and chains of volcanoes on the continents. Subduction zone volcanoes along the edges of deep-sea trenches, such as those in Indonesia and the western Pacific, are among the most explosive in the world, creating new islands and destroying old ones.

The vast majority of volcanism occurs on the bottom of the ocean, where oceanic crust is being created and destroyed. Rifts account for most oceanic volcanism. Along midocean spreading ridges, where the ocean floor is literally spreading apart, magma welling up from the upper mantle forms new oceanic crust and submarine volcanic structures, including volcanoes and basalt flows. The intense volcanism results when the crust weakens and the upper mantle is exposed to the surface. As the pressure on the mantle lowers as it rises toward the surface, the mantle separates into highly mobile magma that erupts basaltic lava, which is highly fluid molten rock. Rift volcanoes on the ocean floor and on the continents erupt massive floods of basalt.

Hot-spot volcanism produces volcanoes in the interiors of plates far from plate margins, where most volcanism occurs. Hot spots are supplied with magma derived from deep within the mantle, possibly from the very top of the core. The magma rises in gigantic mantle plumes, like huge bubbles. They provide a steady flow of molten rock for magma chambers underlying volcanoes.

The best examples of hot-spot volcanism are the volcanoes that built Iceland, the Galápagos Islands, and the Hawaiian Islands. The Hawaiian chain apparently grew from a single mantle plume beginning about 5 million years ago. The islands appear as though assembled on a conveyor belt, with the Pacific plate traversing northwestward over the hot spot about three inches per year. Kilauea, whose Hawaiian name means "much spreading," continues to erupt (Fig. 22). For the last two decades, its lava flows have covered a total of some 40 square miles and added several hundred acres of new land to Hawaii's coast. Kilauea, an infant by geologic standards, has yet to build itself into a true mountain. It is simply a plateau perched on the surface of Mauna Loa, the world's largest active volcano.

A single volcanic eruption of an active volcano can produce anywhere from a few cubic yards to as much as five cubic miles or more of volcanic mate-

Figure 22 *A lava flow entering the sea from the March 28, 1955 eruption of Kilauea, Hawaii.*

(Photo by G. A. Macdonald, courtesy of USGS)

rial consisting of lava and hot, solid particles called pyroclastics (Fig. 23) along with volcanic ash, water vapor, and gases. Subduction zone volcanoes, which account for about 80 percent of the active volcanoes (Table 3), produce about 1 billion cubic yards of volcanic material per year. Most of this material is pyroclastics. The majority of subduction zone volcanoes, some 400 in all, dominate the Pacific region. Rift volcanoes, which account for about 15 percent of the active volcanoes, produce about 2.5 billion cubic yards of volcanic material per year. Most of this material is submarine basalt flows. Hot-spot volcanoes, which account for about 5 percent of the active volcanoes, produce about 0.5 billion cubic yards of volcanic material per year. Most of this material is basalt flows on the ocean floor, while pyroclastics and lava flows occur on the continents.

EARTHQUAKES AS DESTROYERS

Earthquakes are the most highly destructive geologic forces. The damage arising from a major temblor is widespread, altering the landscape for thousands of square miles. The Earth's crust constantly readjusts itself, resulting in verti-

TABLE 3   COMPARISON OF TYPES OF VOLCANISM

Characteristic	Subduction	Rift zone	Hot spot
Location	Deep ocean trenches	Midocean ridges	Interior of plates
Percent active volcanoes	80 percent	15 percent	5 percent
Topography	Mountains, island arcs	Submarine ridges	Mountains, geysers
Examples	Andies Mts., Japan Is.	Azores Is., Iceland	Hawaiian Is., Yellowstone
Heat source	Plate friction	Convection currents	Upwelling from core
Magma temperature	Low	High	Low
Magma viscosity	High	Low	Low
Volatile content	High	Low	Low
Silica content	High	Low	Low
Type of eruption	Explosive	Effusive	Both
Volcanic products	Pyroclasts	Lava	Both
Rock type	Rhyolite, Andesite	Basalt	Basalt
Type of cone	Composit	Cinder fissure	Cinder shield

cal and horizontal displacements on the surface associated with fracture zones in the crust. Large earthquakes can produce offsets of several tens of feet in only a few tens of seconds. The rupturing faults can communicate with other long-distant faults, causing earthquakes up to thousands of miles away.

Figure 23 A pyroclastic flow from the July 22, 1980 Mount St. Helens eruption.

(Photo by L. Wilson, courtesy of USGS)

Earthquakes often form tall, steep-banked scarps or cliffs (Fig. 24) and cause massive landslides that scar the countryside. The greatest deformation occurs near thrust faults. In thrust faults, one block overrides another, especially on the hanging wall—the land that rises during an earthquake. Active faults, which are responsible for scarps, rifts, and mountain ranges, crisscross much of the land surface at plate boundaries on the edges of continents and in the continental interiors underlain by ancient rifts.

Most faults occur at the boundaries between crustal plates. The majority of earthquakes occur when huge plates collide or slide past each other. If the plates hang up in so-called stuck spots called asperites, the sudden release generates tremendous seismic energy. The interaction of plates causes rocks to strain and deform. If deformation occurs near the surface, major earthquakes rupture the land (Fig. 25).

Figure 24 *Fault scarp from the August 1959 Hebgen Lake earthquake, Gallatin County, Montana.*

(Photo by I. J. Witkind, courtesy of USGS)

The most powerful earthquakes occur when one plate thrusts under another in deep subduction zones. The greatest seismic energy is released along the circum-Pacific belt, a band of subduction zones flanking the Pacific Basin. Oceanic and continental rifts, such as the East African Rift with its imposing highlands, impressive escarpments, and dramatic valleys, produce strong earthquakes as well as volcanoes.

Earthquake activity also occurs in stable zones, where strong rocks comprise the shields in the interiors of the continents (refer to chapter 1 about continental shields). Earthquakes in these regions are triggered when the crust is compressed by tectonic forces originating at plate margins, which weakened the crust. The crust could have also been weakened when it was pulled apart along extinct or failed rift systems.

Earthquakes upraise large folds called anticlines that arise from a series of discrete tectonic events over periods of millions of years. Unlike earth-

Figure 25 *Landslide fissures in Chickasaw Bluff, east of Reel Foot Lake, Obion County, Tennessee.*

(Photo by Fuller, courtesy of USGS)

quakes that break along faults, those associated with folds do not rupture the Earth's surface. Many of the world's major fold belts responsible for raising mountain ranges, such as those bordering the Mediterranean Sea, are earthquake prone. The August 26, 1999 Turkey earthquake with a magnitude of 7.4 occurred along a San Andreas–like fault system. Most of these earthquakes occur under young anticlines or upfolded strata less than a few million years old. These folded strata are the geologic product of successive earthquakes resulting from compression forces during plate collisions.

Geologists still do not fully understand why seismic energy unleashes violently in some cases and causes little or no earthquake activity in others. An earthquakes's magnitude is proportional to the length and depth of the rupture created by slipping plates. Generally, the deeper and longer the fault, the larger the earthquake. Other processes that affect earthquake magnitude include the frictional strength of the fault, the drop in stress across the fault, and the speed of the rupture as it traverses over the fault. A break along a fault can travel at speeds of up to a mile a second.

The size of the geographic area affected by earthquakes depends on the earthquake magnitude and the dissipation of seismic energy with increasing distance from the epicenter. Some types of ground transmit seismic energy more effectively than others. For a given magnitude, seismic waves extend over a much wider area in the eastern United States than in the West. This results from a substantial difference in the crustal composition and structure of the two regions. The East consists of older sedimentary strata that carry seismic energy over long distances. In contrast, West contains young igneous and sedimentary rocks sliced up by faults, where seismic energy travels shorter distances.

EARTH MOVEMENTS

The downslope movement of earth materials under the force of gravity plays a substantial role in landform development. Earth movements are classified under the general categories of slides or falls, mass wasting, and liquefaction. Landslides (Fig. 26) are violent movements of earth materials mainly caused by earthquakes, volcanoes, and severe weather. They are among the most powerful erosional agents on land and under the sea. Slides are also triggered by the removal of lateral support resulting from river, glacial, and wave erosion.

Landslides are often large and destructive, as millions of tons of rock derived from a mass of bedrock break into many fragments during the fall. The material behaves as a fluid that spreads out into the valley floor below, possibly running some distance uphill on the opposite side of the valley. Most landslides are rapid movements of overburden that often includes the underlying bedrock. Slides consisting of overburden alone are called debris slides, and

those containing both overburden and bedrock are called rockslides and slumps. Rockslides develop when planes of weakness, such as bedding planes or joining, lie parallel to a slope, especially when undercut by erosion.

Rockfalls involving earth materials dropping from a nearly vertical mountainside are particularly hazardous to highways in mountainous terrain, especially after a heavy downpour. Rockfalls range from individual blocks tumbling down a mountain slope to the catastrophic failure of huge masses weighing hundreds of thousands of tons falling nearly straight down a mountain face. The debris usually builds a loose pile of angular blocks at the base of a cliff called a talus cone (Fig. 27). If large blocks of rock drop into a body of water, they can generate immensely destructive tsunamis that break onto nearby shores.

Slumps develop where a strong, resistant rock overlies weak beds. The material slides down in a curved plane, tilting the resistant member upward,

Figure 26 A landslide that extends over the Pacific Coast Highway, the Pacific Palisades area, Los Angeles County, California on March 31, 1958.

(Photo by J. T. McGill, courtesy of USGS)

Figure 27 *A talus cone on the west wall of Canyon Creek Valley, Glacier National Park, Glacier County, Montana.*

(Photo by H. E. Malde, courtesy of USGS)

while the weaker rock flows out, piling into a heap. Unlike rockslides, slumps develop new cliffs almost as tall as those prior to the slumping, which sets the stage for renewed slumping. Therefore, slumping is a continuous process. Many generations of slumps can often be seen lying far in front of the present cliffs.

Soil slides are the failure of fine-grained, weakly cemented materials on steep slopes. These materials form a wave of sediment sweeping downward at a high rate of speed. Precipitation frees dirt and rocks by raising the water pressure in pore spaces within the soil. As the water table rises and pore pressure increases, friction holding the topsoil layer to the hillside decreases until the material can no longer resist the pull of gravity.

The second type of earth movement, called mass wasting, is the slipping, sliding, and creeping of material down even the gentlest slopes. It is directly influenced by the soil water content. Creep (Fig. 28) is a slow downslope movement of soil. One can recognize it by the results—poles, fence posts, and trees lean downhill. Creep might occur more rapidly in areas where freeze-thaw cycles cause material to move downslope as the soil expands and contracts.

Earthflows occur as the soil water content increases, which adds weight and reduces the stability of a slope by lowering the resistance. It is a transition between the slow and rapid varieties of mass wasting and a more visible form of movement. An earthflow usually has a spoon-shaped sliding surface, upon which a tongue of soil flows a short distance. It leaves a large curved scarp where the material breaks away from the hillside.

Mudflows (Fig. 29) result from a further increase in soil water content. They are highly viscous or thick fluids with the consistency of wet concrete and often carry a tumbling mass of rocks and boulders. Mudflows are the most impressive features of many deserts. Heavy rainfall in bordering mountain regions produces sheets of water that rapidly run down steep mountain slopes, accumulating heavy amounts of sediment along the way. The swift flood of

Figure 28 *Creep in shales, Lancaster County, Pennsylvania.*

(Photo by C. D. Walcott, courtesy of USGS)

muddy material often has a steep wall-like front and can cause serious damage as it flows out of the mountains.

Lahars are mudflows initiated by volcanic eruptions. They are masses of water-saturated rock debris flowing down the steep slopes of a volcano. The volcanic debris consists of masses of loose, unstable rock deposited on the volcano's flanks by explosive eruptions. The water is supplied by rainfall, melting snow, a crater lake, or a nearby reservoir. When crossing a snowfield, a pyroclastic or lava flow rapidly melts it, producing lahars along with floods.

The third type of earth movement is liquefaction. This causes ground failures during earthquakes and violent volcanic eruptions when subterranean sediments liquefy. Liquefaction usually occurs in areas where sands and silts were deposited since the last ice age and where the groundwater table lies near the surface. Generally, the younger and looser the sediments and the shallower the water table, the more susceptible the soil is to liquefaction.

Liquefaction causes lateral spreads, flow failures, and the loss of bearing strength. It also enhances ground settlement or subsidence. Lateral spreads (Fig. 30) are the sideways movement of large blocks of soil, resulting from liquefaction in a subsurface layer during earthquakes. They generally develop on gentle slopes and involve several feet of horizontal movement. However, where slopes are particularly favorable and the duration of ground shaking is lengthy, lateral movement can be quite extensive. Lateral spreads usually break up internally, forming fissures and scarps that mar the landscape.

Flow failures commonly occur in low-lying areas such as tidal flats. They are the most catastrophic type of ground motion associated with liquefaction. Movement of liquefying sediments near the surface causes high ground to sink and spread and causes low ground to rise and narrow. Flow failures consist of soil or blocks of intact material riding on a layer of liquefied soil. They generally move several tens of feet. However, under certain geographic conditions, they can travel several miles at high rates. Flow failures usually form in loose, saturated sands or silts on steep slopes and originate both on land and undersea.

Figure 30 *Slumping and fissuring in western Chimbote, Peru due to liquefaction and lateral spreading of water-saturated sediments.*

(Courtesy of USGS)

EROSION

The most impressive landforms are those carved out of the crust by powerful erosional processes. In the Earth's early stages, erosion rates were probably higher and the relief of the land was not nearly as great as it is today. Millions of years of mountain building and erosion have provided spectacular landscapes of tall mountains and deep canyons.

Ancient geologic structures have long been erased by active erosional agents. Nature's hydrologic cycle, involving the flow of water from the ocean, across the land, and back to the sea, provides the most effective forces of erosion. Massive glaciers carved out some of the most monumental landforms (Fig. 31), while outwash streams from glacial meltwater further eroded the landscape. Areas lacking rainfall or snowfall, such as deserts and tundra, retain much of their geologic structures simply because they experience little erosional activity.

Figure 31 Dome Glacier cascading from the Columbia Ice Fields, showing the moraine-covered tongue of the Athabaska Glacier and abandoned lateral and recessional moraines, Alberta, Canada.

(Photo by F. O. Jones, courtesy of USGS)

Erosion levels rugged mountains, gouges deep ravines into the hardest rock, and obliterates most geologic features on Earth, including ancient man-made structures. No matter how pervasive the formation of mountain ranges by the forces of uplift, mountains ultimately lose the battle with erosion and are worn down, leaving only their deep roots to mark where they once stood.

Rising mountain ranges are matched by erosion, resulting in little net growth. As mountains grow taller, erosion increases due to greater rain, wind, and glacial activity, thereby reducing the growth rate. The cores of the world's mountains contain some of the oldest rocks. These rocks were once buried deep within the bowels of the Earth and were thrust upward by the collision of crustal plates. Huge blocks of crystalline rock in the interiors of the continents were raised by tectonic forces operating deep inside the Earth and exposed by erosion. Alpine glaciers aggressively attack mountains, becoming the world's most potent erosional agents.

Erosion rates vary depending on the amount of precipitation, the topography of the land, the type of rock and soil material, and the vegetative cover. Soil erosion causes the most widespread degradation of the land. It is most effective in regions with little or no plant cover that are subjected to sudden downpours. Falling rain erodes surface material by impact and runoff. About 90 percent of the energy dissipates by the impact. The high-velocity impact of raindrops striking the ground loosens material and splashes it upward. On hillsides, some material falls back farther downslope. Most of the impact splashes are up to a foot high. The lateral splash movement is several times the height, depending on the degree of slope.

Impact erosion accounts for the puzzling removal of soil from hilltops where little runoff occurs. It also ruins soil by splashing up the light clay particles, which are carried away by runoff, leaving behind infertile silt and sand. The runoff flows down the hillside and erodes the soil, cutting deep gullies into the terrain (Fig. 32). The degree of erosion depends on the soil composition, the slope steepness, and the presence of vegetation.

Massive sandstone cliffs that have been slowly eroding in the western United States originated from sediments laid down millions of years ago in a vast inland sea. Perhaps nowhere is this process more telling than in the Grand Canyon of northwestern Arizona. It was excavated over millions of years by the roaring Colorado River, now a mere trickle of its former self.

SEDIMENTATION

Sedimentary processes are slowly at work on the bottom of the ocean. Marine sediments consist of material washed off the continents. Most sedimentary rocks originated along continental margins and in the basins of inland seas that invaded the interiors of North and South America, Europe, and Asia during the Mesozoic era. Areas with high sedimentation rates form deposits thousands of feet thick. When exposed to the surface, individual sedimentary beds can be traced over long distances.

Sedimentary rock begins when erosion wears down mountain ranges and rivers carry the debris to the sea. The sediments originate from the weathering of surface rocks. The rocks weather, or break down, into sediment grains by the action of rain, wind, glacial ice, cycles of freezing and thawing, and plant and animal activities.

Weathering splits rocks apart (Fig. 33) or causes the outer layers to peel or spall off by a process called exfoliation. The products of weathering include a wide range of materials from fine-grained sediments to huge boulders. Exposed rocks on the surface mechanically break down into silts, sands, and gravels and chemically break down into clays and carbonates.

Figure 32 *Gullying of the upper Miocene beds, Kern River district, Kern County, California.*

(Photo by R. W. Pack, courtesy of USGS)

Figure 33 A frost-split glacial boulder in Wallace Canyon, Sequoia National Park, Tulare County, California.

(Photo by F. E. Mattes, courtesy of USGS)

Erosion by rain, wind, or glacial ice supplies rivers with loose sediment grains that are carried downstream to the sea. Angular sediment grains indicate a short transit time. In contrast, rounded sediment grains show severe abrasion by long-distance travel, by reworking of fast-flowing streams, or by pounding of waves on the beach. Indeed, many sandstone formations were once beach deposits that were deeply buried, cemented into hard rock, and exposed on dry land.

Each year, stream runoff transports roughly 40 billion tons of sediment to the ocean, where it settles onto the continental shelf. The towering landform of the Himalayas is the single greatest source of sediment in the world. Rivers draining the region, including the Ganges and Brahmaputra, discharge about 40 percent of the world's total sedimentary load into the Bay of Bengal. Sediment layers accumulate into a 1,000-mile-wide and 3-mile-thick mound, the largest sediment pile in the world.

Rivers transport enormous quantities of sediment derived from their respective continental interiors. The Amazon of South America, the world's largest river, conveys one-sixth of the Earth's running water some 4,000 miles.

Large-scale deforestation and severe soil erosion at its headwaters in recent years has forced the river to carry heavier sediment loads.

The Mississippi River and its tributaries drain a major portion of the central United States from the Rockies to the Appalachian Mountains. Annually, the Mississippi River dumps over a quarter-billion tons of sediment into the Gulf of Mexico, widening the Mississippi Delta and the Gulf coastal region from East Texas to the Florida Panhandle. The area was built up by sediments eroded from the interior of the continent and hauled in by the Mississippi and other rivers. Streams that are heavily laden with sediments overflow their beds. This forces the streams to take several detours as they meander toward the sea (Fig. 34).

The following chapters will take a closer look as some of the major landforms.

Figure 34 *Meandering and oxbow lakes in northern Seward Peninsula region, Alaska.*

(Photo by D. M. Hopkins, courtesy of USGS)

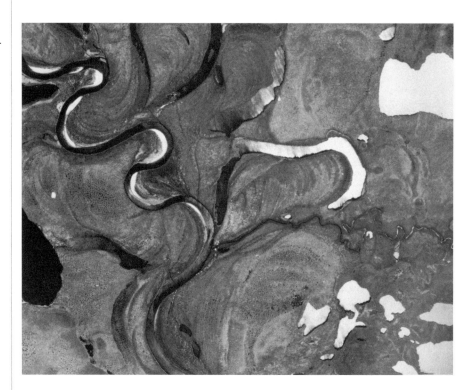

3

MOUNTAIN RANGES
LANDFORMS CREATED BY CRUSTAL UPLIFT

Mountains are by far the most significant landforms in terms of their rugged beauty and importance in controlling the climate and the flow of rivers. This chapter examines how mountains form and discusses the different types of mountains.

The most awe-inspiring landforms are mountain ranges, created by the forces of uplift and erosion. Mountains are areas of high relief in contrast with the rest of the land, rising abruptly above the surrounding terrain (Table 4 and Fig. 35). They are towering topographical features involving massive deformation of the rocks in their cores. The interiors of mountains contain some of the oldest rocks, which were once buried deep in the crust and subsequently thrust to the surface by tectonic activity.

Mountains occur mostly in ranges. Although a few isolated peaks do exist, they are rare. Mountains contain complex internal structures formed by folding, faulting, volcanism, igneous intrusion, and metamorphism. Powerful tectonic forces originating deep within the Earth raised huge blocks of rock during continental collisions. As they struggle upward, mountains ultimately lose the battle with erosion and crumble to the sea.

TABLE 4   THE TALLEST MOUNTAIN PEAKS BY STATE

State	Mountain or peak	Elevation in feet	State	Mountain or peak	Elevation in feet
Alabama	Cheaha Mt.	2,407	Montana	Granite Pk.	12,799
Alaska	McKinley Mt.	20,320	Nebraska	Kimball Co.	5,246
Arizona	Humphreys Pk.	12,633	Nevada	Boundary Pk.	13,143
Arkansas	Magazine Mt.	2,753	New Hampshire	Washington Mt.	6,288
California	Whitney Mt.	14,494	New Jersey	High Point	1,803
Colorado	Elbert Mt.	14,433	New Mexico	Wheeler Pk.	13,161
Connecticut	Frissell Mt.	2,380	New York	Marcy Mt.	5,344
Delaware	Ebright Rd.	442	North Carolina	Mitchell	6,684
Florida	Walton Co.	345	North Dakota	White Butte	3,506
Georgia	Brasstown Bald	4,784	Ohio	Campbell Hill	1,550
Hawaii	Mauna Kea Mt.	13,796	Oklahoma	Black Mesa	4,973
Idaho	Borah Pk.	12,662	Oregon	Mt. Hood	11,239
Illinois	Charles Mound	1,235	Pennsylvania	Davis Mt.	3,213
Indiana	Wayne Co.	1,257	Rhode Island	Jerimoth Hill	812
Iowa	Osceola Co.	1,670	South Carolina	Sessafras Mt.	3,560
Kansas	Sunflower Mt.	4,039	South Dakota	Harney Pk.	7,242
Kentucky	Black Mt.	4,145	Tennessee	Clingmans Dome	6,643
Louisiana	Driskill Mt.	535	Texas	Guadalupe Pk.	8,749
Maine	Katahdin Mt.	5,268	Utah	Kings Pk.	13,528
Maryland	Backbone Mt.	3,360	Vermont	Mansfield Mt.	4,393
Massachusetts	Greylock Mt.	3,491	Virginia	Rogers Mt.	5,729
Michigan	Curwood Mt.	1,980	Washington	Mt. Rainer	14,410
Minnesota	Eagle Mt.	2,301	West Virginia	Spruce Knob	4,863
Mississippi	Woodall Mt.	806	Wisconsin	Timms Hill	1,951
Missouri	Taum Sauk Mt.	1,772	Wyoming	Gannett Pk.	13,804

HISTORIC MOUNTAIN RANGES

During the late Precambrian between 1.3 and 0.9 billion years ago, Laurentia, the ancient North American continent, collided with another large landmass on its southern and eastern borders. This occurred during the creation of the supercontinent Rodinia. The collision raised a 3,000-mile-long mountain belt

in eastern North America during the Grenville orogeny (episode of mountain building). A similar mountain belt occupied parts of western Europe.

Between 630 and 570 million years ago, Rodinia rifted apart. The separated continents dispersed around an ancestral Atlantic Ocean called the Iapetus Sea. Near the present Appalachians is a long belt of volcanism that provides evidence for the breakup of the supercontinent. When all continents reached their maximum dispersal roughly 480 million years ago, subduction of the ocean floor beneath the North American plate initiated a period of volcanic activity and intense mountain building.

At the beginning of the Paleozoic era, continental collisions assembled the southern continents into Gondwana, named for a geologic province in east central India. A major mountain-building episode deformed the margins of all Gondwanan continents (Fig. 36), indicating their collision during this interval. When Laurentia collided with Gondwana about 500 million years ago, the crumpled crust created an ancestral Appalachian Range that continued into western South America long before the formation of the Andes Mountains.

Between 420 million and 380 million years ago, Laurentia collided with Baltica, the ancient European landmass. The collision fused the two continents into Laurasia, named for the Laurentian province of Canada and the Eurasian

Figure 35 *Silverton below the San Juan Mountains, San Juan County, Colorado.*

(Photo by Ransome, courtesy of USGS)

Figure 36 *Late Precambrian orogeny in Gondwana shown in darkened area.*

continent. The closing of the Iapetus Sea as Laurentia collided with Baltica resulted in the great Caledonian orogeny, which created a mountain belt that extended from southern Wales, continued into Scotland, and stretched across Scandinavia and Greenland (Fig. 37).

The Caledonian orogeny continued into North America. It produced a mountain belt from Alabama through Newfoundland that reached as far west as Wisconsin and Iowa. During the Taconian disturbance, extensive volcanism erupted in this region, culminating in a chain of folded mountains that formed the Taconic Range of eastern New York State. Vermont still preserves the roots of these ancient mountains, which heaved upward between 470 and 400 million years ago when Laurasia and the African continent collided.

From 400 million to 350 years ago, a collision between present eastern North America and northwestern Europe raised the Acadian Mountains. The terrestrial redbeds of the Catskills in the Appalachian Mountains of southwestern New York State to Virginia are the main expression of this orogeny in North America. Accompanying the mountain-building episode at its climax was extensive igneous activity and metamorphism.

The Appalachian Mountains (Fig. 38), stretching 2,000 miles from central Alabama to Newfoundland, and the Ouachita Mountains in Arkansas and Oklahoma arose during continental collisions between North America, Eurasia, and Africa during the assembly of Pangaea from 360 million to 270 million years ago. This episode of mountain building also uplifted the Hercynian Mountains in Europe, progressing from England to Ireland and continuing through France and Germany. The folding and faulting was accompanied by large-scale igneous activity in Britain and on the European continent. The Appalachians entered old age when North America separated from Africa, tearing apart the lithosphere upon which the range stood. As a result, the once-towering mountains collapsed.

The Ural Mountains, which resemble no other mountain chain, formed during a collision between the Siberian and Russian Shields between 600 million and 300 million years ago. This occurred at roughly the same time the North American plate collided with the African plate during the building of the Appalachian Mountains. The nearly 2,000-mile-long Ural Mountain Range reaches unprecedented depths of 100 or more miles to the base of the lithosphere underlying central Russia. Unlike most other mountain chains, which have evolved through a similar pattern of growth and collapse, the

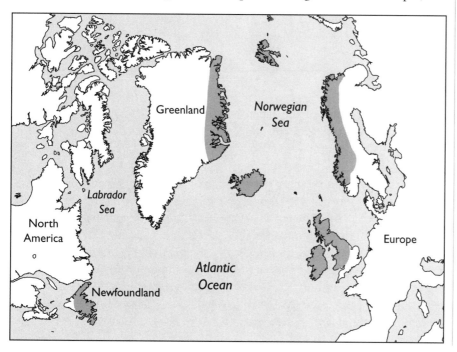

Figure 37 The Caledonian orogeny in Great Britain, Scandinavia, Greenland, and North America (shown in darkened area.)

Figure 38 *The Appalachian Mountains of North America.*

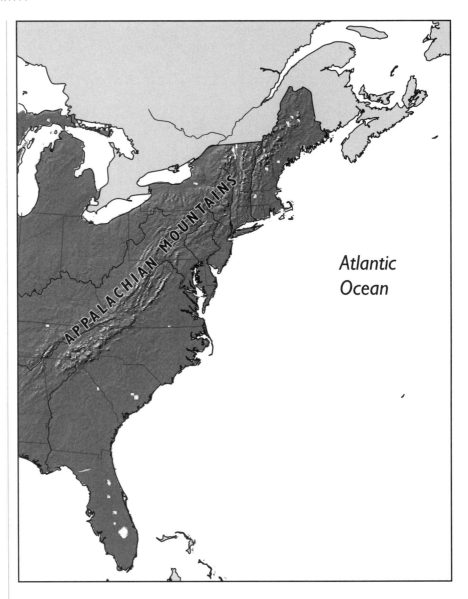

Urals are frozen in midevolution. The mountains have remained intact because Asia never separated from Europe.

The Transantarctic Range, with its great belts of folded rocks, was raised when two plates came together to create the continent of Antarctica (Fig. 39). Prior to the end of the Permian period, the younger parts of West Antarctica had not yet formed and only East Antarctica was in existence. Then more than 60 million years ago, two plates came together to construct the crust of West Antarctica.

The Innuitian orogeny deformed the northern margin of North America by colliding with another crustal plate. The Old Red Sandstone, a thick sequence of chiefly nonmarine sediments in Great Britain and northwest Europe, is the main expression of this mountain-building episode in Europe. The formation contains great masses of sand and mud that accumulated in basins between the ranges of the Caledonian Mountains.

A collision of island arcs with the western margin of North America resulted in a mountain-building event called the Antler orogeny. The island arcs appear to have formed approximately 470 million years ago off the west coast of North America. The orogeny intensely deformed rocks in the Great Basin region from the California-Nevada border to Idaho. The continued clashing of island arcs with North America initiated an episode of mountain building in Nevada called the Sonoma orogeny, which coincided with the final assembly of Pangaea about 250 million years ago.

The mountain belts of the Cordilleran of North America, the Andean of South America, and the Tethyan of Africa-Eurasia arose during plate collisions; one plate slipped under the leading edge of another, thus increasing crustal buoyancy. The Cordilleran and Andean Belts were created by the collision of

Figure 39 One of the mountains in the Darwin Glacier Mountain Range, Antarctica.

(Photo by F. R. Frank, courtesy of U.S. Navy)

East Pacific plates with the continental margins of the newly formed American plates when Pangaea rifted apart about 170 million years ago. The Tethyan Belt formed when Africa collided with Eurasia around 30 million years ago, raising the Alps and Dolomites of Europe. Huge chunks of rock high in the Alps apparently originated hundreds of miles underground, bringing some of the Earth's deepest secrets to the surface.

Between 80 million and 40 million years ago, the Laramide orogeny uplifted a large part of western North America, raising the Rocky Mountain region (Fig. 40) from northern Mexico to Canada nearly a mile above sea level. The mountain building resulted from increased buoyancy of the continental crust by the subduction of oceanic crust and its attached lithosphere beneath North America. Apparently, the subduction was largely horizontal. The oceanic lithosphere slid along the base of the continent as far inland as the Black Hills of South Dakota. The Canadian Rockies were erected by slices of sedimentary rock that were successively detached from the underlying basement rock and thrust eastward on top of each other.

Figure 40 *The Rocky Mountains, looking southwest over Loveland Pass, north of South Park, Colorado.*

(Photo by T. S. Lovering, courtesy of USGS)

A region between the Sierra Nevada Range in California and the southern Rockies uplifted abruptly during the past 20 million years, raising the area over 3,000 feet. The Nevadan orogeny was a surge of volcanic and plutonic activity caused by the subduction of oceanic crust under western North America. It created the Sierra Nevada, which rose an amazing 7,000 feet during the last 10 million years, apparently from an increase in buoyancy by a mass of hot rock in the mantle.

The Indian subcontinent detached from Gondwana about 130 million years ago. Along with Australia riding on the same plate, it sped northward across the ancestral Indian Ocean. Around 45 million years ago, India slammed into southern Asia, making initial contact possibly 10 million years earlier. As the Indian and Asian plates collided, the oceanic lithosphere between them thrust under Tibet, destroying 6,000 miles of subducting plate.

The increased buoyancy raised the Himalaya Mountains (Fig. 41). They stand on a thick shield of strong Precambrian rock, which makes them the tallest range on Earth, comprising 10 of the world's highest peaks. The orogeny also uplifted the broad, high-rising Tibetan Plateau, whose equal has not existed on this planet for over a billion years. Virtually the entire Tibetan Plateau is underlain by Indian lithosphere, the horizontal thrusting of which was responsible for the plateau's vast upheaval.

During the collision between India and Asia, Indochina slid southeastward relative to south China along the 600-mile-long Red River Fault, running from Tibet to the South China Sea. As India plowed into Asia, it pushed Indochina eastward at least 300 miles. In the process of continental escape, Indochina jutted out to sea, rearranging the entire face of southeast Asia. When the fault locked around 20 million years ago, the continental escape halted. The additional stress on Asia thickened the crust and raised the Himalayas (Fig. 42) and the Tibetan Plateau, the largest expanse of land above three miles elevation. In fact, half of the plateau appears to have arisen within the last 10 million years.

About 30 million years ago, an equatorial sea called the Tethys, which separated Africa and Eurasia, began to close off. Thick sediments accumulating on the ocean floor for tens of millions of years buckled into long belts of mountain ranges on the northern and southern flanks (Fig. 43). The contact between the continents initiated the Alpine orogeny, raising the Pyrenees on the border between Spain and France, the Atlas Mountains of northwest Africa, and the Carpathians in east-central Europe. The Alps of northern Italy formed in a similar manner as the Himalayas, when the Italian prong of the African plate thrust into the European plate. However, because the European plate is only half as thick as the Indian plate, the Alps achieved only about half the height of the Himalayas. Having learned of the history of mountain building, the next section examines how mountains were created.

Figure 41 *The collision of the Indian subcontinent with Asia uplifted the Himalaya Mountains.*

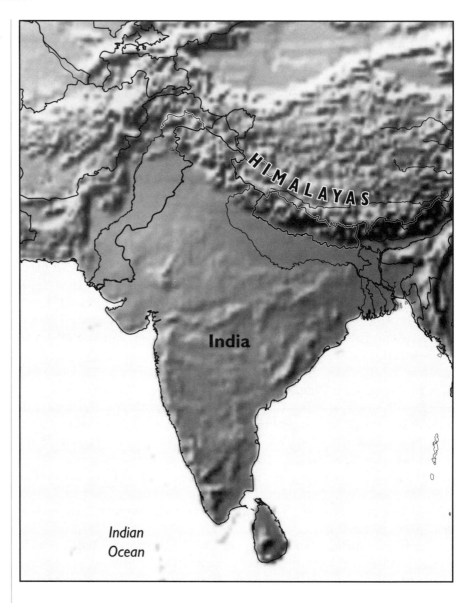

Figure 41 *The collision of the Indian subcontinent with Asia uplifted the Himalaya Mountains.*

MOUNTAIN BUILDING

Prior to the introduction of the plate tectonics theory, the understanding of mountain building was rather rudimentary. Geologists generally thought mountains formed early in geologic history when the molten crust solidified and shriveled, giving the Earth the appearance of a wrinkled baked apple. After making more extensive studies of mountain ranges, however, geologists concluded that the folding of rock layers was far too intense (Fig. 44), requiring considerably more rapid cooling and contraction than was possible. Furthermore, if mountains had formed in this manner, they would have been scattered evenly throughout the world instead of concentrating in a few mountain ranges.

The usual method of building mountains is for plate motions to shove crustal material onto a strong plate. Mountains also form when a deep root of light crustal rock literally floats the range like an iceberg on the high seas. The buoyant continental crust is generally 25 to 30 miles thick. However, under mountains it is an additional 20 miles or more in thickness. Because continental crust is about 80 to 85 percent as dense as the underlying mantle, deep crustal roots can support mountains several miles high.

The underlying lithosphere might drip away from the crust to be replaced by hot rock from the mantle, supplying additional buoyancy that further lifts the mountain range. Globs of relatively cold rock dropping hundreds of miles into the mantle appear to precede this type of mountain building. An

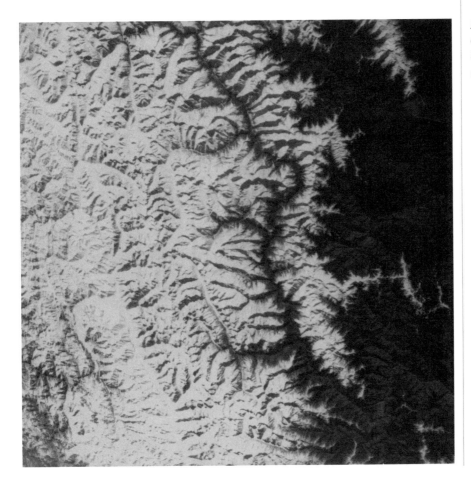

Figure 42 *The south face of the Himalaya Range, India.*

(Courtesy of NASA)

Figure 43 *Mountain belts (darkened areas) formed by the collision of Africa and Eurasia.*

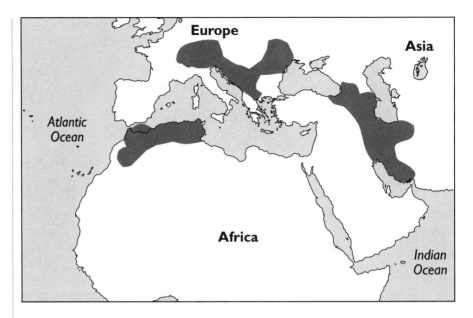

example is the 2.5-mile-high southern Sierra Nevada, which has witnessed a period of rapid growth during the last 10 million years. Interestingly, no plates have converged near the region for more than 70 million years.

Continental collisions during tectonic activity compress the crust (Fig. 45), pushing up huge masses of rock into several mountain ranges. The sutures joining the landmasses are still visible as the eroded cores of ancient mountains called orogens. The long-lived continental roots underlying mountain ranges can extend 100 miles or more downward into the upper mantle. The squeezing of a plate into a thicker one by continental collision might be the very process that forms deep roots.

The mountainous spine of the Andes Range, which runs 4,500 miles along the western edge of South America, arose from an increase in crustal buoyancy by the subduction of oceanic crust beneath the continental plate. A gravity survey high in the Andes indicated less gravitational attraction than that observed at sea level. This implies that the granites in the core of the mountains were less dense than the rocks lying below. The study was the first to prove that continents comprised lighter granitelike materials and the ocean floor consisted of heavier basaltlike substances. The difference in density between the two rock types buoyed the continents, a relationship known as *isostasy*, which comes from Greek and means "equal standing."

Seismologists are scientists who study the propagation of seismic waves though the Earth. While attempting to solve a long-standing puzzle about the origin of the southern Andes, seismologists employed marine seismic survey techniques in the inland waterways that weave through the southern portion of the mountains in the Patagonia region. A ship that bounced seismic waves, which are like sound waves, off the ocean floor and received them from a string of hydrophones towed behind the vessel detected irregularities in the crust, producing a sonic picture of the mountain's core. The aim was to determine why the rapidly rising Andes are so much larger than expected from buoyant forces alone.

The Himalaya Mountains provide the key to understanding mountain building. The complex tectonic underpinning of the Tibetan Plateau has increased buoyancy and uplifted the entire region. Additional crustal material underthrusts the submerged crust, and the increased buoyancy raises mountain ranges. Additional compression and deformation farther inland beyond the line of collision produces a high plateau with surface volcanoes, similar to Tibet.

Figure 44 *Tilted beds between Rawlins and Laramie, Wyoming.*

(Photo by J. R. Balsley, courtesy of USGS)

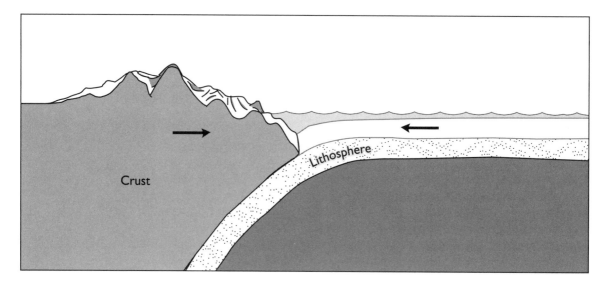

Figure 45 *Mountain ranges are formed by the collision of tectonic plates.*

The strain of raising the Himalayas by the collision of the Indian plate with the Eurasian plate has resulted in deformation and earthquakes all along the contact between the two blocks. In September 1993, a 6.1 magnitude quake leveled 20 villages in India, killing 10,000 people. An enormous seismic zone stretches across Tibet and much of China. During the 20th century, more than a dozen earthquakes of magnitude 8.0 or more have struck the region. India is still plowing into Asia. However, as resisting forces continue to build, plate convergence will eventually cease. At that time, the mountains will stop growing, only to be toppled by erosion.

The shaping of mountains depends as much on the destructive forces of erosion as on the constructive power of plate tectonics. The interactions between tectonic, climatic, and erosional processes exert strong control over the shape and height of mountains as well as the amount of time needed to build or destroy them. Erosion might actually be the most powerful agent of mountain building. It removes mass that is restored by isostasy, which lifts the entire mountain range to replace the missing mass. If the rate of erosion matches the rate of uplift, the size and shape of mountains can remain stable for upward of millions of years. As the mountains age, the crust supporting them thins out, and erosion takes over to bring down even the most imposing ranges.

FOLDED MOUNTAINS

Folded mountain belts, where strata have buckled over, are created by the collision of continental plates. They constitute a massive deformation of rocks in the core of the range (Fig. 46). Plate motions built most mountains by shoving the crust of one plate onto another. A gently sloping fault beneath the

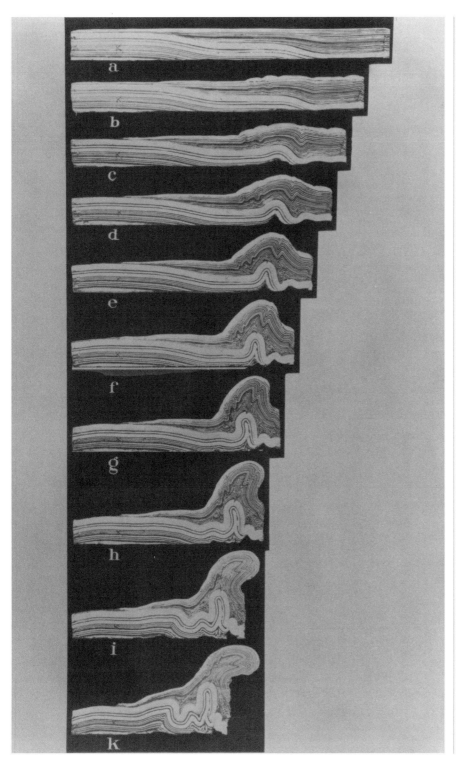

Figure 46 *A wax model illustrating mountain uplift.*

(Photo by J. K. Hillers, courtesy of USGS)

Wind River Mountains of Wyoming suggests that horizontal squeezing of the continents rather than vertical lifting formed these as well as many other mountain ranges. Mountain building, which provides the forces that fold and fault rocks at shallow depths, also generates stresses that strongly distort rocks deep below.

The increased weight of thousands of feet of sediments deposited in trenches along the seaward margin of a continental plate presses downward on the oceanic crust, causing it to bulge. As continental and oceanic plates merge, the heavier oceanic plate slides under the lighter continental plate, forcing it deep into the mantle. The compression of the sedimentary layers of both plates causes the leading edge of the continental crust to swell.

As subduction continues, the topmost layers are scraped off the descending oceanic crust and are plastered against the swollen edge of the continental crust. The addition of crustal material produces a mountain belt similar to the Andes. This particular range owes its existence to an increase in crustal buoyancy by the subduction of the Nazca plate beneath the South American plate along the Chile trench at a rate of about two inches per year (Fig. 47). The interaction of the two plates accounts for the crumpling of the stable continental margin to form belts of folded mountains that constitute the eastern ranges. The western portions of the Andes comprise a chain of active volcanoes and enormous batholiths. The two ranges are separated by a high plateau called the Altiplano, the tallest part of the Andes.

Figure 47 *The Andes were raised by increased crustal thickness as the Nazca plate plunges under the South American plate.*

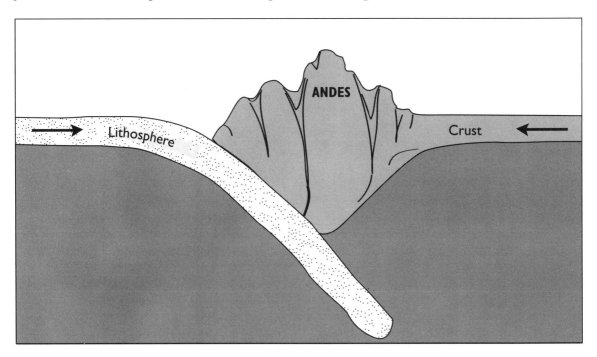

Continental collisions arising from plate tectonics stabilize a portion of the mobile mantle rock beneath the continents. As a result, the continents drift along with thick layers of chemically distinct rock that is as much as 250 miles thick. These rocks underlying the continents are called subcrustal keels. The process that forms deep roots operates by squeezing a plate during continental collisions, thus thickening the plate. This activity was most intense during the collision between the Indian plate and the Eurasian plate, which has shrunk some 1,000 miles while raising the great Himalayas and the Tibetan Plateau, the largest tableland in the world.

Water and wind gradually erase almost all signs of once splendid mountain ranges. However, erosion does not remove everything. Often, the roots of ancient mountains, such as the Taconic Range in eastern New York, survive beneath otherwise unimpressive terrain. Buried deep down are huge faults running through the basement rock, indicating that long ago, tectonic forces squeezed the crust into mountains now completely leveled.

Similar folds and faults form the roots of present ranges such as the Appalachian Mountains. The rocks under the northwest Appalachians are riddled with ancient fractures called Iapetan faults that are being forced into earthquake activity by the compression of the crust. The timing of the squeezing, heating, and alteration of rock within mountains is unknown. Yet according to analysis of radioactive elements used in dating rocks, the process of mountain folding apparently spans only a few million years.

When continents collide, they crumple the crust and force mountain ranges up at the point of impact. Many of today's folded mountains were uplifted by Paleozoic continental collisions that raised huge masses of rocks into several mountain belts throughout the world (Fig. 48). The Appalachians were once towering mountains that have now eroded down to mere stumps. These mountains formed when North America, Eurasia, and Africa slammed into each other during the late Paleozoic formation of Pangaea. The main uplift culminated between 350 million and 250 million years ago.

Sedimentary layers lying beneath the core of the Appalachians suggest that thrusting involving basement rocks causes the formation of folded mountain belts. Data also suggest that this process of mountain building has occurred since plate tectonics began more than 2.7 billion years ago. The shoving and stacking of thrust sheets during continental collision also appears to have been an important mechanism in the continued growth of the continents.

Underlying the southern Appalachians are over 10 miles of essentially undeformed sedimentary and metamorphic rocks. The surface rocks, by comparison, were highly deformed by the continental collision that formed Pangaea. This primarily occurred from thrust faulting, which involved crustal material being carried horizontally for great distances. The sedimentary strata riding westward on top of Precambrian metamorphic rocks folded over, buck-

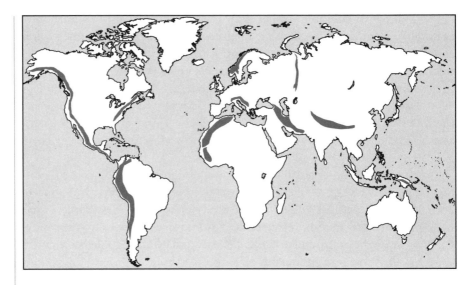

ling the crust into a series of ridges and valleys. Similarly, the Mauritanide mountain range on the opposite continent in western Africa was built simultaneously during the collision with North America and mimics the Appalachians in many respects.

CRYSTALLINE MOUNTAINS

The first rocks to form on Earth originated directly from molten magma and are therefore called igneous, from the Latin meaning "fire." Most igneous rocks were derived from new material in the mantle, some began when oceanic crust subducted into the mantle, and others formed by the melting of continental crust. The first two processes continuously build the continents, whereas the last adds little to the total volume of continental crust. Pockets of molten rock provide new source material for volcanoes and other igneous activity. Highly active continental movements are accompanied by intense volcanism, which also affects the rate of mountain building.

Igneous rocks are classified by their mineral content and crystalline texture, which are governed by the degree of separation and rate of cooling of the magma. The most common crystalline rocks are Precambrian granites and metamorphics (Fig. 49). These constitute most of the interiors of the continents. The rate of cooling controls the texture of granitic rocks. Long cooling periods yield the largest crystals, and short cooling periods result in significantly smaller grains.

Large crystals form late in the crystallization of a magma body such as a batholith, the most massive of igneous bodies. These huge granitic structures add significant amounts of new crust to a continent. (The best-known batholiths include the Sierra Nevada, the California, and the Andean.) Large crystals also grow in a smaller volume in the presence of volatiles such as water

Figure 49 *The contact between Manitou lime-stone and Precambrian gneiss and schist, Priest Canyon, Fremont County, Colorado.*

(Photo by J. C. Maher, courtesy of USGS)

and carbon dioxide. As the magma body slowly cools, the crystals grow directly out of the fluid melt or from the volatile magmatic fluids that invade the surrounding rocks.

Heat is the primary agent for recrystallization. Rocks must often be buried deep inside the Earth to generate the temperatures and pressures needed for extensive metamorphism. In the deepest part of the continents, where temperatures and pressures are extremely high, rocks partially melt and metamorphose. Varying degrees of metamorphism are also achieved at shallower depths in geologically active areas with higher thermal gradients, where the temperature increases with depth at a much faster rate than normal. During metamorphism, rocks behave plastically. They bend or stretch due to the high temperatures and pressures exerted by the overlying strata. Their abundance make metamorphics the predominant rock types of the continental crust.

Magma bodies that invade the crust assimilate the surrounding rocks as they melt toward the surface. This results in two major classes of igneous rocks. *Intrusives* are derived from the invasion of the crust by a magma body. *Extrusives* are derived from the eruption of magma onto the surface. Because their source materials are essentially the same, both rock types share similar chemical compositions. However, they have different textures due to the difference in cooling rates.

Intrusive magma bodies cool over long periods—perhaps a million years or more—because the rocks they invade are good insulators that tend to hold in the heat. The magma body is thus able to segregate into various components, allowing large crystals to grow. Extremely large crystals form *pegmatites,*

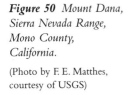

Figure 50 *Mount Dana, Sierra Nevada Range, Mono County, California.*

(Photo by F. E. Matthes, courtesy of USGS)

Figure 51 *Fremont Canyon with Henry Mountains in the background, Wayne and Garfield counties, Utah.*

(Photo by J. R. Stacy, courtesy of USGS)

which comes from Greek and means "fastened together." Generally, the larger the magma body the longer the cooling period and, consequently, the larger the crystals.

Plutons are intrusive magma bodies with a variety of shapes and sizes. The most massive plutons are batholiths, which are usually elongated and exceed 40 square miles of surface exposure. Batholiths consist of granitic rocks with large crystals, comprising mostly quartz, feldspar, and mica. They produce some of the world's major mountain ranges. Examples include California's Sierra Nevada (Fig. 50), which is nearly 400 miles long but only about 50 miles wide, and the Idaho Batholith, which is 250 miles long and 100 miles wide.

A stock is a smaller intrusive magma body shaped similar to a batholith but having less than 40 square miles of surface exposure. It might also be a projection of a larger batholith lying below. Like a batholith, a stock comprises coarse-grained granitic rocks and forms smaller mountain ranges.

A dike is an intrusive magma body that is tabular in shape and considerably longer than wide. Dikes form when magma fluids occupy large cracks or fissures in the crust. Because dike rocks are usually more resistant than the surrounding material, they generally form long ridges when exposed by erosion. Sills are similar to dikes in their tabular shape but form parallel to planes of weakness, such as sedimentary beds. A laccolith is a type of sill that bulges the overlying sediments upward, often creating isolated mountain peaks such as the Henry Mountains in southern Utah (Fig. 51). Prominent laccolithic domes are responsible for several other mountainous areas in the mile-high Colorado Plateau region, where doming occurred some 25 million years ago.

FAULT-BLOCK MOUNTAINS

To the west of the Rocky Mountains, parts of which originated by block faulting, many parallel faults sliced through the Basin and Range Province between the Sierra Nevada of California and the Wasatch Mountains of Utah. They resulted in a series of north-south trending fault-block mountain ranges. The crust in this region is bounded by faults and literally broken into hundreds of steeply tilted blocks that rise nearly a mile above the basin. These produce nearly parallel mountain ranges up to 50 miles long. They formed by the stretching of the crust, creating some of the thinnest crust on any continent.

Figure 52 *The Wasatch Range, north-central Utah.*

(Photo by R. R. Woolley, courtesy of USGS)

Figure 53 *The Grand Teton Range near Moran Junction, Teton County, Wyoming.*

(Photo by I. J. Witkind, courtesy of USGS)

The Sierra Nevada and the Cascades to the north, were forced southwestward to make room for crustal expansion.

Death Valley, at 280 feet below sea level, once stood several thousand feet higher. The region collapsed when the continental crust thinned from extensive block faulting. The Great Basin area is a remnant of a broad belt of mountains and high plateaus that subsequently collapsed after the crust pulled apart during the rising of the Rocky Mountains. The still-rising Wasatch Range (Fig. 52) in north-central Utah and south Idaho is an excellent example of a north-trending series of faults, one below the other. The fault blocks extend 80 miles, with a probable net slip along the west side of 18,000 feet. The Grand Tetons of western Wyoming (Fig. 53) were upfaulted along the eastern flank and downfaulted to the west. The rest of the Rocky Mountains evolved by a process of upthrusting connected with plate collision and subduction. The Rockies' stability might result from their position over a relatively cool part of the mantle, which allows them to deform only very slowly over time.

About 30 million years ago, the North American continent approached the East Pacific Rise spreading ridge system. The first portion of the continent to override the axis of seafloor spreading was the coast of southern California and northwest Mexico. As the rift system and subduction zone converged, the intervening oceanic plate dove into a deep trench. Sediments caught in the trench were compressed and thrust upward to form California's Coast Ranges. A system of faults associated with the 650-mile-long San Andreas Fault (Fig. 54 and Fig. 55) crisscross the mountain belt. The Sierra Nevada Range to the east has risen about 1.3 miles over the last 10 million years and is apparently buoyed by a mass of hot rock in the upper mantle.

The faults of the Mojave and adjacent Death Valley absorb about 10 percent of the total slippage between the Pacific and North American plates. The complex crustal movements associated with these faults are responsible for most of California's tectonic and geologic features such as the Coast Ranges and the Sierra Nevada. Moreover, these faults generate most of the earthquakes that plague the nation's shakiest state.

The next chapter will discuss another type of mountain formed by volcanic activity.

Figure 54 *A view northwest toward San Francisco, California, along the San Andreas Fault.*

(Photo by R. E. Wallace, courtesy of USGS)

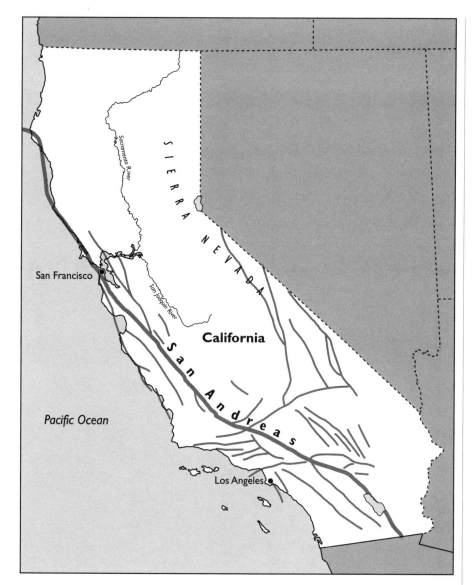

Figure 55 The San Andreas and associated faults in California.

4

VOLCANIC TERRAIN
LANDFORMS CREATED BY VOLCANISM

Volcanoes are the most important land formers. They produce numerous geologic structures, ranging from a variety of cones to huge lava flows. Volcanoes can be very destructive. A single eruption can often wipe out entire towns and take the lives of thousands of people. Volcanoes also significantly impact the climate. This chapter examines volcanoes and their effects on the landscape.

More than three-quarters of the Earth's surface above and beneath the sea are of volcanic origin. Volcanoes, whose eruptions are beneficial as well as hazardous, are the most spectacular of all of Earth's processes. Their destructibility is evident by the devastation they have wreaked on civilization down through the ages. They also play a constructive role by adding material to the crust, thereby building entirely new landscapes.

Undersea volcanic eruptions produce oceanic crust by the shifting of lithospheric plates, comprising the crust and underlying lithosphere, which move the continents around. Volcanism on the ocean floor is responsible for erecting most of the world's islands. Successive eruptions pile up volcanic rocks until the volcanoes finally break through the surface of the sea. Because these volcanoes rise from the very bottom of the ocean, they make the tallest mountains in the world.

THE RING OF FIRE

An almost continuous ring of fire surrounds the Pacific Basin. It coincides with the circum-Pacific belt, which is known for its extensive seismic activity. This is because the same tectonic processes that generate earthquakes also produce volcanoes. These volcanoes are associated with a band of subduction zones around the Pacific Rim. Moreover, almost all mountain ranges surrounding the Pacific Ocean formed by plate subduction. As oceanic crust subducts into the mantle, it melts to provide molten magma for volcanoes fringing the deep-sea trenches. Consequently, most of the 600 active volcanoes in the world lie in the Pacific Ocean, with nearly half residing in the western Pacific region alone (Fig. 56).

The vast majority of the world's volcanoes accompany crustal movements at plate margins where lithospheric plates are either diverging or converging. Fissure eruptions on the ocean floor occur at the boundaries between lithospheric plates. The brittle crust pulls apart by the process of seafloor spreading, which creates new oceanic crust. Basaltic magma welling up along the entire length of a fissure forms large lava pools similar to those on Hawaii's most active volcano Kilauea.

Most volcanoes occur on plate boundaries associated with deep-sea trenches along the margins of continents and are adjacent to island arcs fringing subduction zones. The zones also form arcs because this is the geometric pattern formed when a plane cuts a sphere. Subduction zone volcanoes build volcanic chains on continents when a lithospheric plate dives into a subduction zone and slides beneath the continental crust. As the lithospheric plate subducts into the hot mantle, portions of the descending plate along with the

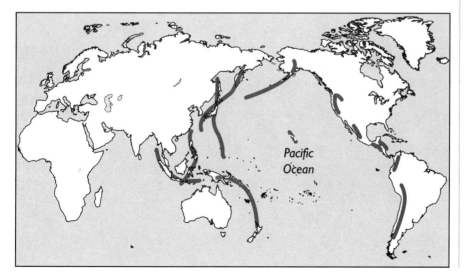

Figure 56 *Major belts of active volcanoes.*

adjacent crustal plate melt under the high temperatures, forming pockets of magma. The molten rock rises toward the surface to feed magma chambers underlying active volcanoes.

Volcanic islands begin as undersea volcanoes. At subduction zones where one plate descends beneath another, magma forms when the lighter constituents of the subducted oceanic crust melt. The upwelling magma creates island arcs, mostly in the Pacific. These islands include Indonesia, the Philippines, Japan, the Kuril Islands, and the Aleutians, the world's longest arc, extending more than 3,000 miles from Asia to Alaska.

Beginning at the western tip of the Aleutian Islands, a string of volcanoes stretches along the Aleutian Archipelago. They consist of a series of volcanic islands formed by the subduction of the Pacific plate into the Aleutian Trench. The band of volcanoes turns south across the Cascade Range of British Columbia, Washington, Oregon, and northern California. The volcanic

Figure 57 *Total devastation of vegetation on the near side of the cone due to the May 18, 1980 eruption of Mount St. Helens, Skamania County, Washington.*

(Courtesy of USGS)

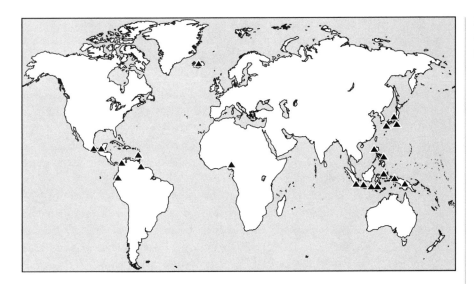

activity in these mountains originates from the subduction of the Juan de Fuca plate into the Cascadia subduction zone. The May 18, 1980 lateral blast from Mount St. Helens caused total destruction along its flanks (Fig. 57), providing a vivid reminder of the explosive nature of subduction zone volcanoes.

The chain of volcanoes continues along Baja California and southwest Mexico and on through western Central America. Many active cones have erupted in this part of Central America in recent years, including Nevado del Ruiz of Columbia. Its massive mudflows killed 25,000 people in November 1985. It is grouped with some 20 other volcanoes considered the most dangerous in the world (Fig. 58).

The band of volcanoes journeys along the course of the Andes Mountains on the western edge of South America, which are known for their highly explosive nature from the subduction of the Nazca plate into the Chilean Trench. The volcanic belt then turns toward Antarctica and the islands of New Zealand that sit on the margin of the Pacific and Australian plates, New Guinea in the Malay archipelago, and Indonesia. In Indonesia, the volcanoes Tambora and Krakatau have produced the greatest eruptions in modern history as a result of the subduction of the Australian plate into the Java Trench.

The volcanic belt continues across the Philippines. The subduction of the Pacific plate into the Philippine Trench resulted in the June 1991 Pinatubo eruption, whose huge ash cloud dramatically changed the world's climate. From there, the zone of volcanoes runs across the Japanese Islands. These islands were built from a combination of rock drawn from the mantle along a volcanic arc and from sediments scraped off the ocean floor. The volcanic belt finally ends on the Kamchatka Peninsula in northeast Asia, whose volcanoes are renowned for their powerful blasts.

VOLCANIC STRUCTURES

Volcanoes eject a variety of rock types. They range from rhyolite to basalt. Rhyolite is the lightest of volcanic rocks, with a high silica content. Basalt is the heaviest of volcanic rocks, with a low silica and high iron-magnesium content (Table 5). Basalt is the most common igneous rock produced by the extrusion of magma onto the surface. Most of the over 500 active volcanoes in the world are completely or predominantly basaltic. A remarkable type of volcano called a kimberlite pipe rises from great depths within the mantle and brings to the surface diamonds formed some 3 billion years ago.

Volcanic cones display a variety of shapes and sizes, depending on the composition of the erupted magma. The four main types of volcanoes are cinder cones, composite volcanoes, shield volcanoes, and lava domes. Cinder cones are the simplest volcanic structures, built from particles and congealed lava ejected from a single vent. Explosive eruptions form short, steep slopes usually less than 1,000 feet high. Cinder cones build upward and outward by accumulating layers of pumice, ash, and other volcanic debris falling back onto the volcano's flanks. The general order of events is eruption, followed by formation of cone and crater, and then lava flow. The bowl-shaped craters are numerous in western North America. Mexico's Paricutin Volcano, which erupted in a farmer's field in 1943, is a classic example (Fig. 59).

Composite volcanoes are built up from both cinder and lava cemented into tall mountains rising several thousand feet. They are generally steep sided

---

TABLE 5   CLASSIFICATION OF VOLCANIC ROCKS

Property	Basalt	Andesite	Rhyolite
Silica content	Lowest, about 50%, a basic rock	Intermediate, about 60%	Highest more than 65%, an acidic rock
Dark-mineral content	Highest	Intermediate	Lowest
Typical minerals	Feldspar Pyroxene Olivine Oxides	Feldspar Amphibole Pyroxine Mica	Feldspar Quartz Mica Amphibole
Density	Highest	Intermediate	Lowest
Melting point	Highest	Intermediate	Lowest
Molten rock viscosity at the surface	Lowest	Intermediate	Highest
Formation of lavas	Highest	Intermediate	Lowest
Formation of pyroclastics	Lowest	Intermediate	Highest

Figure 59 *The cinder cone of Paricutin Volcano, Michoacan, Mexico.*

(Photo by W. F. Foshag, courtesy of USGS)

and comprise symmetrical lava flows, volcanic ash, cinders, and blocks. The crater at the summit contains a central vent or a cluster of vents. During eruption, the hardened plug in the volcano's throat breaks apart by the buildup of pressure from trapped gases below. Molten rock and fragments shoot high into the air and fall back onto the volcano's flanks as cinder and ash.

Layers of lava from milder eruptions reinforce the fragments, forming cones with a steep summit and steeply sloping flanks (Fig. 60). Lava also flows through breaks in the crater wall or from fissures on the flanks of the cone, continually building it upward. As a result, composite volcanoes are the tallest cones in the world. They often end in catastrophic collapse, thus preventing them from becoming the highest mountains. Sometimes, after a volcano becomes dormant, erosion eats away at the cone until only the volcanic plug remains.

Shield volcanoes are extremely broad. The slope along the flanks rise only a few degrees and no more than 10 degrees near the summit. They erupt, for the most part, basaltic lava from central vents. Highly fluid molten rock oozes from the vent or violently squirts out, forming fiery fountains of lava from pools within the crater. As the lava builds in the center, it flows outward in all directions, forming a structure similar to an upside-down dinner plate. The lava spreads out and covers large areas, as much as 1,000 square miles. Several of these dome-shaped features in northern California and Oregon, such as the Mono-Inyo Craters, are 3 or 4 miles wide and 1,500 to 2,000 feet high. Hawaii's Mauna Loa is the largest active shield volcano, creating a huge, sloping dome rising 13,675 feet above sea level, much less than its depth below the sea.

Lava domes grow by expansion from within because the lava is too viscous or heavy to flow very far. As a result, it piles up around the vent. Lava domes commonly occur in a piggyback fashion within the craters of large, composite volcanoes. Good examples are California's Mono Domes and Lassen Peak (Fig. 61).

When a volcano explosively erupts fluid lava high into the air, the volcanic ash disperses by the wind. The eruption produces particles ranging in size from ash to molten blobs of lava up to several feet wide called volcanic bombs (Fig. 62). Often, they change shape while flying through the air and flatten or splatter when striking the ground. If the bombs are the size of a nut, they are called lapilli, which is from the Latin for "little stones." These form strange, gravel-like deposits along the countryside. As volcanic bombs cool in flight, they assume a variety of forms called cannonball, spindle, bread-crust, cow dung, ribbon, or fusiform, depending on the shape or surface appearance. Bread-crust bombs, which can reach several feet across, are named for their crusty appearance caused by gases escaping from the bomb during the hardening of the outer surface. Some volcanic bombs actually explode when landing due to the rapid expansion of gases in the molten interior as the solid crust cracks open on impact. Often, volcanic bombs gyrate wildly through the air, sometimes whistling like incoming artillery fire.

Figure 61 Devastated area below Lassen Peak, Shasta County, California.

(Photo by W. G. Pierce, courtesy of USGS)

Figure 62 *A volcanic bomb weathered out of the Vega Bay Formation, Aleutian region, Alaska.*

(Photo by R. R. Coats, courtesy of USGS)

All solid particles ejected into the atmosphere from volcanic eruptions are collectively called tephra, from the Greek meaning "ash." Tephra includes an assortment of fragments ranging from large blocks to dust-size material. It originates from molten rock containing dissolved gases. The magma rises through a conduit and suddenly separates into liquid and bubbles when nearing the surface. With decreasing pressure, the bubbles grow larger.

If this event occurs near the orifice, a mass of froth spills out and flows down the sides of the volcano, forming pumice. If the reaction occurs deep down in the throat, the bubbles expand explosively and burst the surrounding liquid, which fractures the magma into fragments. Like pellets from a shotgun, the fragments are driven upward by the force of the rapid expansion and hurled high above the volcano. The fragments cool and solidify during their flight through the air. Fine particles are caught by the wind blowing across the eruption cloud and are carried long distances, with ash dropping on the ground along the way.

Nearly all volcanoes produce some tephra. Even relatively quiet eruptions occasionally eject fountains of highly fluid lava. Its spray solidifies into a minor amount of tephra that is confined to the neighborhood of the vent. If water from the sea, a lake, or a water table enters the magma chamber, it instantaneously flashes into steam. This causes violent explosions to rise through the conduit accompanied by little or no new magma. Most tephra produced in this manner originates from the walls of the conduit or from shattered parts of the crater.

Tephra supported by hot gases caused by a lateral blast of volcanic material is called nuée ardente, which is from the French for "glowing cloud." The cloud of ash and pyroclastics flows streamlike near the ground; it might follow existing river valleys for tens of miles at speeds upward of 100 miles per hour. The best-known example was the 1902 eruption of Mt. Pelée, Martinique. In minutes, it annihilated 30,000 inhabitants. When the tehpra cools and solidifies, it forms deposits called ash-flow tuffs that can cover an area up to 1,000 square miles or more.

Lava is molten rock or magma that manages to reach the throat of a volcano or fissure vent without exploding into fragments and flows freely onto the surface. The magma that forms lava is much less viscous than that which produces tephra. This allows volatiles and gases to escape with comparative ease and gives rise to much quieter and milder eruptions.

The outpourings of lava come in two general classes, which take Hawaiian names and are typical of Hawaiian eruptions. They are pahoehoe (pronounced pah-HOE-ay-hoe-ay), which means satinlike, and aa (pronounced AH-ah), which is the sound of pain when walking over them barefoot. Pahoehoe or ropy lavas are highly fluid basalt flows produced when the surface of the flow congeals, forming a thin, plastic skin. The melt beneath continues to flow, molding and remolding the skin into billowing or ropy-looking surfaces. When the lavas eventually solidify, the skin retains the appearance of the flow pressures exerted on it from below.

Aa or blocky lava forms when viscous, subfluid lavas press forward, carrying along a thick and brittle crust. As the lava flows, it stresses the overriding crust, breaking it into rough, jagged blocks. These are pushed ahead of or dragged along with the flow in a disorganized mass (Fig. 63).

After a stream of lava forms a crust and hardens on the surface and the underlying magma continues to flow away, it creates a long cavern or tunnel called a lava tube (Fig. 64). This tube can reach several feet across and extend for hundreds of feet. The walls and roof of the lava tube are occasionally adorned with stalactites, and the floor is covered with stalagmites composed of lava deposits.

In the volcanic terrain of Alaska, Washington, Oregon, California, and Hawaii, the collapse of shallow tunnels in lava flows forms deep depressions.

Figure 63 *A lava flow, Hawaiian Volcanoes National Park, Hawaii.*

(Courtesy of USGS)

Long caverns beneath the surface of a lava flow are created by the withdrawal of lava as the surface hardens. In exceptional cases, they can extend up to several miles within a lava flow. Often, circular or elliptical depressions on the surface of the lava flows result from the collapse of the lava tube roofs. One example is a large, collapsed depression in a lava flow in New Mexico that is nearly a mile long and 300 feet wide.

THE WORLD'S HOT SPOTS

Scattered throughout the world are more than 100 small regions of isolated volcanic activity known as hot spots. The Hawaiian Islands were produced by such a hot spot. Unlike most active volcanoes, those created by hot spots rarely exist at plate boundaries but lie deep in the interiors of plates. They are notable for their isolation. They are far removed from normal centers of volcanic and earthquake activity. They might be the only distinguishable feature of an otherwise unimpressive landscape.

Hot spots provide a pipeline for transporting heat from the planet's interior to the surface. The magma plumes rise through the mantle as separate giant bubbles of hot rock. When a plume passes the boundary between the lower and upper mantle, some 400 miles below the surface, the bulbous head separates from the tail and rises. This is often followed millions of years later by another similarly created plume, producing a one-two volcanic punch.

Figure 64 *Entrance to Thurston lava tube in First Twin Crater, Halemaumau Volcano, Hawaii County, Hawaii.*

(Photo by H. T. Stearns, courtesy of USGS)

Vast floods of basalt were produced by hot-spot volcanism as plumes of magma rose to the surface from deep within the mantle. For example, India's Deccan Traps are counted among the greatest outpourings of basalt. They erupted about 65 million years ago when large volumes of molten magma poured onto the surface from a giant rift running down the west side of the subcontinent. Over a period of several million years, about 100 separate flows spilled more than 350,000 cubic miles of lava onto much of west-central India. The lava totaled some 8,000 feet thick.

Massive outpourings of basalt covered Washington, Oregon, and Idaho, creating the Columbia River Plateau. The Columbia River Basalt group consists of about 300 lava flows (Fig. 65) that erupted between 17 million and 6 million years ago. The floods of lava enveloped an area of about 65,000 square miles. In places, the lava reached 10,000 feet thick. Periodically, volcanic eruptions spewed batches of basalt as large as 1,200 cubic miles, forming lava lakes up to 450 miles wide in only a matter of days.

When the mantle plumes reach the oceanic crust, they build a succession of islands as the crustal plate slides over them. The ascending plumes can lift entire regions. For instance, in the 3,000-mile-wide section of the South Pacific seafloor, where several hot spots have erupted and formed *chains* of Polynesian islands. Similar swells exist under the Hawaiian chain in the central Pacific, Iceland in the North Atlantic, and the Kerguelen Islands in the southern Indian Ocean. The most active hot spots lie beneath the big island of Hawaii and beneath Réunion Island east of Madagascar.

The passage of a crustal plate over a hot spot often leaves a trail of volcanoes. That trend reveals the direction of plate motion. The volcanic structures generally align obliquely to the adjacent midocean ridge system. The hot spot's track might be a continuous volcanic ridge or a chain of islands and seamounts rising high above the seafloor.

The most prominent and easily recognized hot spot fueled the five separate volcanoes that created the Hawaiian Islands about 5 million years ago. Hawaii's Mauna Kea Volcano, which built most of the main island, is—in effect—the world's tallest mountain. It rises about 33,000 feet above the ocean floor, exceeding the height of Mount Everest by nearly 4,000 feet. Mauna Loa, the largest shield volcano, formed by flow upon flow of 24,000 cubic miles of lava, making it the most voluminous mountain on Earth.

Lying some 3,000 feet below sea level, about 20 miles south of Hawaii, is a submerged volcano named Loihi. Perhaps in another 50,000 years from now, the volcano will finally poke through the surface of the ocean and take its place as the newest member of the Hawaiian chain. The rest of the Hawaiian Islands are progressively older, with extinct volcanoes trailing off to the northwest, followed by the submerged volcanoes of the Emperor Seamounts. Each island erupted in succession as though riding on a conveyer

Figure 65 *Regional joint or fault in Columbia River Basalt, lower Snake River Project, Washington.*

(Photo by F. O. Jones, courtesy of USGS)

belt, with the Pacific plate traversing about 3.5 inches per year over the hot spot (Fig. 66).

Nearly all hot-spot volcanism occurs in regions of broad crustal uplift or swelling where magma lies near the surface. More than half of all hot spots exist under continents. When a continental plate hovers over a number of hot

Figure 66 *The Hawaiian Islands formed by a Pacific Plate drifting over a single hot spot.*

spots, molten magma welling up from deep below creates domelike structures in the crust. The growing domes develop deep fissures through which magma rises to the surface. They average about 125 miles wide and account for about 10 percent of the Earth's total land area.

The largest concentration of hot spots occur in Africa (Fig. 67), which has remained essentially stationary over the past 30 million years. Hot spots might have been responsible for the unusual topography of the African continent, characterized by numerous basins, swells, and uplifted highlands. Their presence suggests that the African plate has come to rest over a population of hot spots.

Further evidence indicates that Africa has not moved significantly. Hot-spot lavas of several different ages are superimposed on one another. If the continent had been drifting, the hot spot lavas would have been laid down laterally in a chronological sequence. Hot spots are also plentiful in Antarctica and Eurasia, indicating these regions are practically motionless as well. In contrast, on rapidly moving continental plates, such as North and South America, hot spot volcanism is rare.

Hot spots on the continents leave distinguishable trails of volcanoes. The hot spot's track also weakens the crust, cutting through the lithosphere like a hot knife through butter. A hot spot underlying Yellowstone National Park, Wyoming can be traced 400 miles across the Snake River plain in southern Idaho. A rift might eventually develop in this region, where the hot spot's trail has weakened the North American plate. Over the last 15 million years, the North American plate has traveled southwestward over the hot spot, placing the plate under its temporary home at Yellowstone. As the plate continues to drift, the relative motion of the hot spot will bring it across Wyoming and Montana, possibly coming to rest in Canada.

VOLCANIC CALDERAS

A volcanic caldera or large crater, which comes from the Spanish meaning "cauldron," forms when the sudden ejection of large volumes of molten rock

Figure 67 *The Brandberg structure in Namibia near Cape Cross in Southwest Africa is a zone of weakness in the Earth's crust created by upwelling magma.*

(Courtesy of NASA)

from a magma chamber lying a few miles below the surface abruptly removes the underpinning of the chamber's roof. The roof collapses, leaving a deep, broad depression (Fig. 68). The infusion of fresh molten rock into the magma chamber slowly heaves the floor of the caldera upward, providing a vertical uplift approaching several thousand feet.

Calderas generally develop over a mantle plume or hot spot that melts near-surface rocks. They originate in areas where the fractured crust allows easy access for magma to rise to the surface. The intruded magma domes the overlying crust upward, creating a shallow magma chamber that contains a large volume of molten rock. The doming stresses the surface rock that forms the roof of the chamber, causing it to collapse along a ring fracture zone, which becomes the outer wall of the caldera after the eruption.

Many calderas are recognized by widespread secondary volcanic activity such as hot springs and geysers. Yellowstone is such a caldera (Fig. 69). As with all resurgent calderas, Yellowstone formed above a large, persistent mantle plume that melted huge volumes of rock. Three major volcanic episodes have occurred in Yellowstone during the last 2 million years. About 600,000 years ago, a massive eruption disgorged some 250 cubic miles of ash and pumice. The volcanic eruption created the huge Yellowstone Caldera, encompassing an area of about 45 miles long and 25 miles wide. It is a typical resurgent caldera, whose floor has slowly domed upward at an average rate of about three-quarters of an inch per year for the last 70 or so years.

Two other calderas have erupted in the United States over the last 2 million years. About 1.6 million and again 1.2 million years ago, massive volcanic outbursts created the 15-mile-wide Valles Caldera in northern New Mexico. California's Long Valley Caldera is a 2-mile-deep depression measuring about

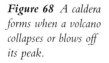

Figure 68 *A caldera forms when a volcano collapses or blows off its peak.*

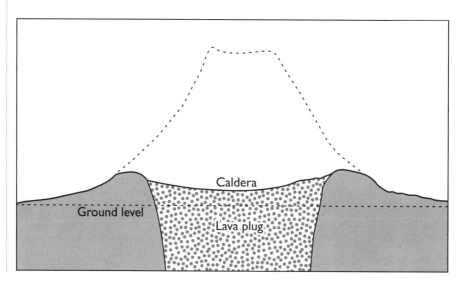

20 miles long and 10 miles wide. It resulted from a cataclysmic eruption 760,000 years ago. The volcano fragmented the nearby mountains into rocky debris. About 120 cubic miles of volcanic material were strewn over a wide area, traveling as far as the Atlantic Coast. Mammoth Mountain is a young volcano within the caldera that has experienced an extended period of activity and appears poised for another eruption at any time.

The Long Valley Caldera lying in east Yosemite National Park encompasses features such as Devils Postpile National Monument and Mammoth Lakes (Fig. 70) and is growing about an inch per year. Many other calderas no more than 20 to 30 million years old lie in a broad belt in the Basin and Range Province of Nevada, Arizona, Utah, and New Mexico. They generally occupy zones where the crust is thinning and the mantle rises near the surface. Perhaps the largest caldera in North America up to 45 miles wide is La Garita in the San Juan Mountains of southwestern Colorado, which erupted about 28 million years ago.

Several calderas have erupted in other parts of the world within the last million years. In northern Sumatra, a massive eruption 73,500 years ago created the Toba Caldera, which subsided as much as a mile or more. It is the

Figure 69 *Terrace Mountain and chaotic landslide blocks of Silver Gate, Yellowstone National Park, Wyoming.*

(Photo by K. E. Barger, courtesy of USGS)

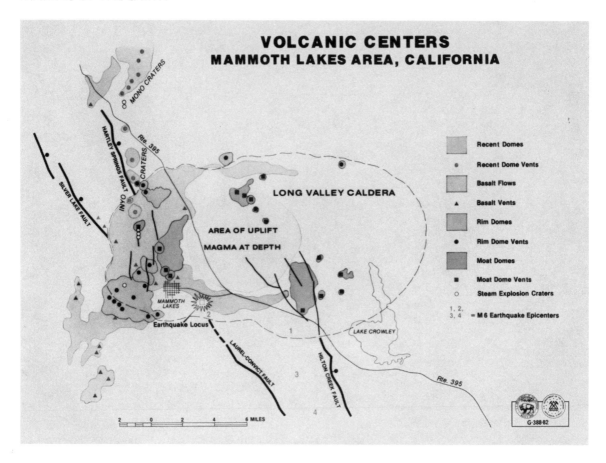

VOLCANIC CENTERS
MAMMOTH LAKES AREA, CALIFORNIA

LONG VALLEY CALDERA

AREA OF UPLIFT

MAGMA AT DEPTH

MAMMOTH
LAKES

Earthquake Locus

LAKE CROWLEY

Rte. 395

	Recent Domes
	Recent Dome Vents
	Basalt Flows
	Basalt Vents
	Rim Domes
	Rim Dome Vents
	Moat Domes
	Moat Dome Vents
	Steam Explosion Craters
1, 2, 3, 4	= M 6 Earthquake Epicenters

2   0   2   4   6 MILES

G-388-82

Figure 70 *Long Valley Caldera, showing volcanic centers of Mammoth Lakes area, California.*

(Courtesy of USGS)

world's largest known resurgent caldera and extends nearly 60 miles in its longest dimension.

A smaller caldera results when a powerful volcano erupts and decapitates itself by blowing off its upper peak, leaving a broad crater generally over a mile across. Often, the lava blows outward, enlarging the crater. The highly viscous lava also might create a plug in the crater that slowly rises, forming a spire or dome (Fig. 71).

CRATER LAKES

Dormant calderas, filled with freshwater from melting snow or rain, form crater lakes. They are among the deepest lakes in the world, depending on the depth of the caldera floor and the water level below the crater rim. Erosion widens the caldera, while sediments eroded from the wall accumulate into thick deposits on the lake bottom. Resurgence of the caldera floor sometimes creates an island capped with young lake sediments.

The world's largest crater lake fills the huge Toba Caldera, created by perhaps the greatest volcanic outburst in the last 2 million years. The caldera formed when the roof over a large magma chamber suddenly collapsed. The floor of the caldera consequently subsided more than a mile and filled with water to form a deep lake. Later, the caldera floor heaved upward several hundred feet like a huge piston, creating Samosir Island, which is 25 miles long, 10 miles wide, and possibly still rising.

Oregon's Crater Lake (Fig. 72) originated when the upper 5,000 feet of the 12,000-foot composite cone of Mount Mazama collapsed about 6,000 years ago and filled with rainwater and melted snow. The lake is 6 miles wide and 2,000 feet deep, the sixth deepest in the world. The rim of the caldera rises 500 to 750 feet above the water's surface. At one end of the lake is a small volcanic peak called Wizard Island, which uplifted during a subsequent period of volcanic activity.

In the Katmai region on the northeast end of the Alaskan Peninsula, a gigantic explosion on June 6, 1912 tore open the bottom of the west slope of Mount Katmai. The top 1,200 feet of the volcano exploded and collapsed into a large caldera 1.5 miles wide and 2,000 feet deep. It later filled with water from melting snow to form a crater lake (Fig. 73).

Crater lakes on Mount Kilauea, Hawaii are filled not with water but, instead, with molten lava. The 2-mile-wide Kilauea Crater is shaped like an inverted saucer rising 4,090 feet above sea level. A large crater forms at the summit, from which two rift zones radiate. The eruptions are usually limited to the crater and the rift zones, particularly the eastern rift zone and the fire pit in the crater.

During the formation of a lava lake, magma from a reservoir erupts to the surface and flows into a depression. A conduit or vent connects the crater

Figure 71 *Lava dome in the crater of Mount St. Helens, Skamania County, Washington.*

(Photo by J. Huges, courtesy of USDA-Forest Service)

to the magma chamber. When fluid magma moves up the pipe, it is stored in the crater until the crater fills and overflows. During periods of inactivity, back flow can completely drain the crater.

Kilauea has erupted on average at least once a year since 1952. Its lava lakes are basalt flows from previous eruptions that have been trapped in large pools. The lake depth can be substantial, as much as 400 feet for the Kilauea Iki Crater. The lakes slowly cool and solidify, generally taking up to a year for shallow lakes and as long as 25 years for the deepest lake at Kilauea Iki. Eventually, the natural dikes that channel the lava into the lake collapse, cutting off the lava flow. The lake then begins to solidify from the bottom up and from the top down. Some lava lakes disappear completely down the bottom of the crater as though the drain plug were pulled.

Lava lakes often erupt vigorous fire fountains that spray lava high into the air (Fig. 74). Mauna Loa is known for its tall fountains of white-hot lava shooting several hundred feet high, forming a spectacular *curtain of fire*. Despite

Figure 72 *Crater Lake, Klammath County, Oregon created by the collapse of Mount Mazama.*

(Photo by H. R. Cornwall, courtesy of USGS)

their violent outbursts, the eruptions are generally harmless and a great delight to the local citizens.

FUMAROLES AND GEYSERS

During the June 1912 Mount Katmai eruption, a massive outpouring of lava created the Valley of Ten Thousand Smokes. A series of explosions excavated a depression at the west base of the volcano, whereupon viscous lava rose pan-cake-fashion 200 feet high and 800 feet in diameter. The entire valley became a hardened, yellowish orange mass, 12 miles long and 3 miles wide. Thousands of volcanic steam vents called fumaroles, which are responsible for giving the valley its name, gushed out of the ground and shot hot-water vapor as high as 1,000 feet into the air.

Figure 73 *The Novarupta Rhyolite Dome is thought to be the main site of the June 16, 1912 Katmai Volcano, Katmai region, Alaska eruption.*

(Courtesy of USGS)

Fumaroles are volcanic vents that expel hot gases. They exist on lava flows, in the calderas and craters of active volcanoes, and in areas where hot, intrusive magma bodies lie near the surface. Fumaroles and geysers are a common sight on the volcanic island of Iceland. Indeed, the term geyser derives from the Icelandic word *geysir,* meaning "gusher." The gas temperature within a fumarole can reach 1,000 decrees Celsius. The gases consist mostly of steam and carbon dioxide along with smaller quantities of nitrogen, carbon monoxide, argon, hydrogen, and other gases. A fumarole called a solfatara, from the Italian meaning "sulfur mine," emits mostly sulfur gases.

A volcanic hot spot beneath Yellowstone is responsible for the continuous thermal activity that spawns numerous fumaroles and geysers such as Old Faithful (Fig. 75). Its nearly hourly eruptions can last several minutes, spouting a column of steam 130 feet high. In addition, a variety of boiling mud pits and

Figure 74 *An alae eruption of Kilauea Volcano, providing a large lava fountain, Hawaii County, Hawaii.*

(Photo by G. A. Smathers, courtesy of National Park Service)

Figure 75 *Eruption of Old Faithful Geyser, Yellowstone National Park, Wyoming.*

(Courtesy of USGS)

hot-water streams are produced when rainwater seeping into the ground acquires heat from a magma chamber and rises through fissures in the torn crust.

Geysers are often intermittent and explosive. Hot water ejected with great force rises 100 to 200 feet high. A column of thundering steam usually follows the jet of water. The tube leading to the surface from a deep underground geyser chamber is often restricted or crooked, similar to the drainpipe under a sink. When water seeps into the chamber from a water table, it is heated from the bottom up. The overlying weight of the water column exerts great pressure onto the water at the bottom of the chamber, keeping it from boiling even though temperatures greatly exceed the normal boiling point of water.

As the water temperature gradually increases, water near the top of the geyser tube boils off, decreasing the weight and causing the water in the bottom of the chamber to flash into steam. The rapid expansion overcomes the restriction. Hot water and superheated steam gush out of the vent to great heights. The record holder is New Zealand's Waimangu Geyser, which in 1904 spouted water and steam 1,500 feet high into the air.

The next chapter will shift attention to rivers and the landforms they build.

5

RIVER COURSES
LANDFORMS CREATED BY FLOWING WATER

Running water is responsible for sculpting the landscape more than any other natural process. This chapter examines how rivers shape the land and discusses the widespread effects of flooding.

Rivers erode valleys and provide a system of drainage delicately balanced with the climate, topography, and lithology. Streams carry off sediments delivered to the stream channel by tributaries and by slope erosion on the sides of the valley. The sediment transported by rivers is then temporarily stored by deposition in the channel and on the adjacent floodplain.

Weathering, downslope movement, and river flow work together to reshape the continents. Even in the most arid regions, the principal topographic features, including mesas and arroyos, are excavated by stream erosion. Rivers carry sediments eroded from the highlands to the sea. Along the way, they carve out new landforms, including deep ravines in rugged terrain and winding valleys in flat terrain. Excess water during floods overflows riverbanks and levees. It also washes over floodplains, replenishing them with new fertile soil.

RIVER BASINS

A river basin is the entire region from which a stream and its tributaries receive water. Each tributary is fed by lesser tributaries down to the smallest rill, which began as a thin film of rainwater called sheet wash. Sediment grains loosened by raindrop impact erosion are carried downstream and eventually dumped into the ocean.

Some 3.5 million miles of rivers and streams flow across the United States. The Mississippi River and its tributaries drain an enormous section of the central portion of the country from the Rockies to the Appalachians (Fig. 76). All tributaries emptying into the Mississippi also have their own drainage areas and together form parts of a larger basin.

The primary purpose of a river is to transport debris eroded at its headwaters and along its banks. The sediment is derived from rocks weathered by rain, wind, and ice. A river continues to erode its bed and build it back up. Erosion and sedimentation therefore determine the shape of a river course. The course can be confined to a single, straight channel when eroding, or it can be meandering or braided when clogged with debris. The sediment-laden stream eventually empties into standing bodies of water (Fig. 77). Worldwide, about 40 billion tons of sediment are carried by stream runoff into the ocean annually.

River load is the type of material a stream carries. The river content consists of suspended load, bed load, and dissolved load. The suspended load is fine material that settles out very slowly and is thus carried long distances. The total amount of sediment in suspension increases downstream as more tributaries enter the river and add their sediment load. The suspended load is near-

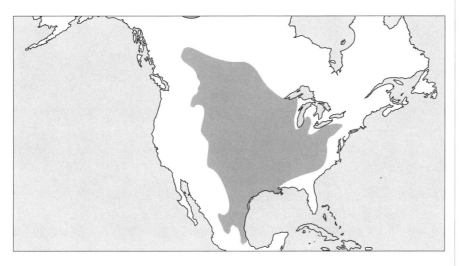

Figure 76 Major river basins (darkened area) in North America.

Figure 77 *Heavy sediment on Lake Tahoe, California and Nevada, from construction activities.*

(Photo by B. Heckathorn and C. Saulisberry, courtesy of USDA-Soil Conservation Service)

ly two-thirds the total river content and amounts to about 25 billion tons per year. The bed load is material such as pebbles and boulders that travel by rolling and sliding along the river bottom during high flow or flood and is a quarter or less of the total river content. The dissolved load is derived from chemical weathering and from solution by the river itself and comprises about 10 percent of the total river content.

Rivers erode by abrasion and solution. Abrasion occurs when the transported material scours the sides and bottom of the channel. A common type of abrasion forms kettle-shaped depressions called potholes. They are worn in riverbeds by rocks spinning around inside the potholes by turbulent eddy currents. Many of the largest potholes formed during the melting of the glaciers after the Ice Age. The impact and drag of the water itself also erode and transport material. Most of the dissolved matter in a stream originates from groundwater draining from a breached water table. Materials such as limestone dissolve in river water that is slightly acidic. Limestone also acts as a buffer to maintain acidity levels within tolerable limits for aquatic life.

The effect of erosion is to deepen, lengthen, and widen river valleys. At the head of a stream, where the slope is steep and water flow is fast, downcutting lengthens the valley by a process called headward erosion. This is mainly how streams cut into the landscape. Farther downstream, both the velocity and discharge increase while the sediment size and the number of banks decrease. These factors allow the river to transport a larger load with lesser slope. Bends in the stream tend to slow the river flow by lowering the gradient.

Erosion widens a stream valley by creep, landsliding, and lateral cutting. The process is most pronounced on the outsides of irregular curves, where the valley side might be undercut by flowing water. Therefore, streams with migrating curves tend to widen their valleys. Many streams have distinctive symmetrical curves called meanders that distribute the river's energy uniformly. Intersecting meanders form isolated cutoff sections of the river called oxbow lakes (Fig. 78).

The rate of erosion depends on the rainfall, evaporation, and vegetative cover in the drainage basin. The ability of a river to erode and transport material depends largely on the velocity, the water flow, the stream gradient, and the shape and roughness of the channel. The average rate of erosion in the United States is about 2.5 inches per 1,000 years. The Columbia River Basin has the lowest erosion rate at 1.5 inches per 1,000 years. The Colorado River Basin has the highest at 6.5 inches per 1,000 years.

Rivers transport enormous quantities of sediment derived from their continental interiors. India's Ganges River carries four times more sediments than the Amazon River, which happens to be three times larger. The Ganges and Brahmaputra Rivers together convey about 40 percent of the world's total amount of sediment discharged into the ocean as erosion gradually wears down the Himalaya Mountains. The mountains' remains are dumped into the Bay of Bengal, creating a massive sediment pile as much as three miles thick.

STREAM DEPOSITS

River deposits called alluvium accumulate due to a decline in stream gradient or inclination, a reduction in stream flow, or a decrease in stream volume; the heaviest material settles out first. Changes in the river environment occur during entrance into standing water, encounters with obstacles, evaporation, and freezing. River deposition is divided into deposits in bodies of water, alluvial fans, and deposits within the stream valley itself. A medium-sized river will take about a million years to move its sandy deposits 100 miles downstream. Along the way, the grains of sand are polished to a high gloss.

Figure 78 *An oxbow lake, Juneau District, southeastern Alaska.*

(Photo by G. K. Gilbert, courtesy of USGS)

River-deposited sedimentary rocks within streambeds are relatively rare because rivers deliver most of their sediment load to lakes or the sea. River deltas develop where rivers enter larger rivers or standing water (Fig. 79 and Fig. 80). The velocity of the river slows so abruptly when entering a body of water that its bed load immediately drops out of suspension. Offshore currents, which create marine or lake deposits, also rework much of the river's load.

Alluvial fans (Fig. 81), which are generally found in arid regions, are similar to river deltas. They form where streams flow out of mountains onto broad valleys. The ground abruptly flattens, causing the stream to slow and deposit its sediment load into a fan-shaped body. As the stream constantly shifts its position, the alluvial fan grows steeper, thicker, and coarser, developing a characteristic cone shape.

Alluvium laid down in the channel of a stream is called channel fill. Accumulations of fill assume many shapes generally known as sand bars. They collect along the edges of a stream, especially on the insides of bends. They also accumulate around obstructions and pile up into submerged shoals and low islands. These deposits are not permanent features but are destroyed, redeposited, or shift positions as conditions on the river change.

Natural levees build up on the banks of a river during a flood. When floodwaters overtop a river channel and flow onto the adjacent floodplain, the

Figure 79 *Delta of the Chelan River entering the Columbia River, Chelan Ferry, Washington.*

(Photo by B. Willis, courtesy of USGS)

Figure 80 The delta of Camerai Creek at the north end of Waterton Lake, Alberta, Canada.

(Photo by C. D. Walcott courtesy of USGS)

velocity quickly diminishes, causing deposition near riverbanks. The levees help keep the river within its banks during normal flow. However, at flood stage, the valley floor is often lower than the river level, inundating the land when flood-waters crest over the levee top. During the 1993 Midwestern floods, perhaps the worst of the 20th century, levee breaks deposited several feet of sand onto farmlands along the swelling rivers. Man-made levees continue to break during major floods, compounding the amount of death and destruction.

When a river excavates part of its floodplain due to changes in flow, it leaves a terrace standing above its new level (Fig. 82). Terraces first develop in the lower reaches of a stream and extend upstream, cutting into sediments laid down earlier in either direction. Terraces also form by the lateral cutting of bedrock by a river. One of the most common causes of terrace building is associated with glacial meltwater during the end of the last ice age. Melting glaciers produced more debris than rivers could handle, and the excess sedi-

ment was deposited in the river valleys. As conditions returned to normal, terraces were formed when rivers downcut into these deposits.

An unusual type of river flow called a braided stream (Fig. 83) forms when the bed load is too large and coarse for the slope and the amount of discharge. The banks easily erode, which chokes the channel with sediment, causing the stream to divide and rejoin repeatedly. The stream deposits the coarser part of its abundant load to attain a steep enough slope to transport the remaining load. This forces the stream to broaden and erode its banks. Alluvium rapidly deposits in constantly shifting positions, forcing the stream to split into interlacing channels that continuously separate and reunite.

DRAINAGE PATTERNS

Streams and their valleys combine into networks that display various types of drainage patterns, depending on the terrain. If the terrain has a uniform com-

Figure 82 Outwash terrace in the Roaring Fork River Valley below Aspen, Pitkin County, Colorado.

(Photo by B. H. Bryant, courtesy of USGS)

Figure 83 *Braided channels of the Nelchina River, Copper River region, Alaska.*

(Photo by J. R. Williams, courtesy of USGS)

position and does not determine the direction of valley growth, the drainage pattern is dendritic (Fig. 84), resembling the branches of a tree. Granitic and horizontally bedded sedimentary rocks generally yield this type of stream pattern.

A trellis drainage pattern displays rectangular shapes that reflect differences in the bedrock's resistance to erosion. The major tributaries of the master stream follow parallel zones of least resistance over underlying folded rocks. Rectangular drainage patterns also occur when fractures crisscross the bedrock, forming zones of weakness particularly susceptible to erosion. If streams radiate outward in all directions from a topographic high such as a volcano or dome, they produce a radial stream drainage pattern.

Drainage patterns are influenced by topographic relief and rock type. They provide important clues about the geologic structure of an area. In addition, the color and texture of the structure impart information about the rock formations that comprise it. Surface expressions such as domes, anticlines, synclines, and folds bear clues about the subsurface structure. Various types of

Figure 84 *A well-developed dendritic drainage pattern on the soft rocks of the Ogallala Formation, Baca County, Colorado.*

(Photo by T. G. McLaughlin, courtesy of USGS)

drainage patterns imply variations in the surface lithology or rock type. The drainage pattern density also indicates the lithology. Variations in the drainage density also correspond to changes in the coarseness of the alluvium.

In areas of exposed bedrock, drainage patterns depend on the lithology of the underlying rocks, the attitude of rock units, and the arrangement and spacing of planes of weakness encountered by runoff. Any abrupt changes in the drainage patterns are particularly important. They signify the boundary between two rock types, which could indicate mineral emplacement.

CANYONS AND VALLEYS

Canyons best exemplify the power of erosion. They generally cut through arid or semiarid regions, where the effect of stream erosion is greater than weath-

Figure 85 *Grand Canyon from Mojave Point.*

(Photo by J. R. Balsley, courtesy of USGS)

ering. Perhaps no better example is the Grand Canyon of northern Arizona (Fig. 85). It is a huge gash in the Earth's crust 277 miles long, up 20 miles wide, and over 1 mile deep. It lies at the southwest end of the Colorado Plateau, a practically mountain-free expanse stretching from Arizona northward into Utah and eastward into Colorado and New Mexico.

Early in the Earth's history, the canyon region was nearly completely flat. Over the last 2 billion years, heat and pressure buckled the land into mountains that were later flattened by erosion. Again, mountains formed. Once more they were eroded and flooded with shallow seas. The land uplifted another time during the growth of the Rocky Mountains and the rising of the Colorado Plateau.

The Grand Canyon is a relatively young feature, formed by uplift and erosion as the Colorado River sliced through a half-billion years of accumulated sediments and Precambrian basement rock. Some 10 to 20 million years ago, the Colorado River began eroding layers of sediment along the Colorado Plateau. Its present course was established between 5 and 6 million years ago when Baja California separated from Mexico, providing a new outlet to the sea. Much of the canyon was not carved out piecemeal, eroding away grain by grain, but by catastrophic landsliding that tore away entire canyon walls.

The canyon provides some of the best rock exposures on the North American continent. It cuts through rock formations hundreds of millions of years old. Well-exposed sedimentary layers give a nearly complete account of the geologic history of the region. On the bottom of the canyon lies the original ancient basement rock, comprising the 1.8-billion-year-old Vishnu Schist, upon which sediments slowly accumulated layer by layer.

Over a mile of sedimentary rocks overlie the dark bedrock of the Grand Canyon. The oldest of these rocks dates to about 540 million years ago, leaving more than a billion years of history missing from the geologic record. During this time, erosion wore down the floor of the Grand Canyon, creating a gap in time known as a hiatus. The rock configuration that includes the hiatus is called an unconformity. The Great Unconformity, readily exposed in the Grand Canyon, stretches across much of North America from Arizona to Wisconsin to Alberta, Canada. Just above the unconformity lies a layer of sandstone that was deposited along an ancient shoreline. The youngest sediments lie near the center of the continent.

Unconformities are common and seen most everywhere. They are among the most remarkable of geologic features and reflect processes that represent the entire spectrum of geologic events. Unconformities contain pockets of time or gaps in the geologic record where no rocks exist. Often, the older, underlying rocks are folded or tilted and overlain by younger, flat-lying sediments. This produces bewildering patterns of discordant rocks and angular contacts. A unique angular unconformity separating the upper horizontal

Plateau series from the older tilted Grand Canyon series adorns the canyon wall (Fig. 86).

As the thick marine sediments slowly deposited on the floor of the Grand Canyon, their continuous buildup caused the ancient seafloor to subside due to the increased weight. In a fraction of the time required to deposit the sediments, a gradual upheaval brought them to their present elevation. During this time, the Colorado River gouged out layer upon layer of rock, exposing hundreds of millions of years of geologic history.

Perhaps less impressive than canyons but just as important to river flow are valleys. A river valley such as the Mississippi Valley is a linear, low-lying track of land traversed by a river or stream and bordered on both sides by higher ground called a floodplain. A narrow valley is not much wider than the river channel itself. In contrast, a broad valley exceeds the width of the river channel many times. A narrow valley is carved out by a fast-flowing river that is actively downcutting in areas of regional uplift, referred to as the youthful stage. Some narrow valleys occur in resistant rocks that slow the lateral cutting of a river and commonly display rapids and waterfalls.

A river widens its valley as it flows along a leveling grade and is no longer rapidly downcutting, referred to as the mature stage. This condition

Figure 86 The upper horizontal Plateau series is separated by an angular unconformity from the older, tilted Grand Canyon series.

(Photo by L. F. Noble, courtesy of USGS)

occurs mostly near the mouth of a river, where wide floodplains exist. Meanders are common features of wide valleys (Fig. 87), especially in areas with uniform banks composed of easily erodible sediment. The valley might widen by flooding, weathering, and mass wasting. Many river valleys were also widened by glaciers during the Ice Age, converting V-shaped valleys into U-shaped ones (Fig. 88).

When rivers clog with sediment and fill their channels, they spill over onto the adjacent plain and carve out a new river course. In the process, they meander downstream, forming thick sediment deposits in broad floodplains that can fill an entire valley. As a stream meanders across a floodplain, the greatest erosion takes place on the outside of the bends, resulting in a steep cut bank in the channel (Fig. 89). On the inside of the bends, however, the water slows and deposits its suspended sediments. During a flood, a winding river often takes a shortcut across a low-lying area separating two bends. This temporarily straightens the river until it further fills its channel with sediment,

Figure 87 *Meanders of Crooked Creek, Mono County, California.*

(Photo by W. T. Lee, courtesy of USGS)

causing it to meander once again. Meanwhile, the cutoff sections of the original river bends become oxbow lakes.

A river also might capture a nearby stream, which is known as piracy. This creates a larger expanse of flowing water. The river grows at the expense of other streams and becomes dominant because it contains more water, erodes softer rocks, or descends a steeper slope. The river therefore has a faster headward erosion that undercuts the divide separating it from another stream and captures the stream's water.

Floodwaters rapidly flowing out of dry mountain regions carry a heavy sediment load, including blocks the size of automobiles. When the stream reaches the adjacent desert, its water rapidly percolates into the desert floor, bringing it to an abrupt halt. Sometimes huge monoliths hauled out of the mountains dot the desert landscape as tributes to the tremendous power of water in motion.

Fluvial, or river, deposits are terrestrial sediments that remain mostly on the continents after erosion. They are recognized in outcrops by their coarse sediment grains and cross-bedding features (Fig. 90), generated when a stream meandered back and forth over old river channels. River currents also align

Figure 88 *Pronounced U shape of the Kern Canyon by glacial erosion, Tulare County, California.*

(Photo by F. E. Matthes, courtesy of USGS)

109

Figure 89 *Severe stream bank erosion along Muddy Creek, Cascade County, Montana.*

(Photo by T. McCabe, courtesy of USDA-Soil Conservation Service)

mineral grains and fossils, giving rocks a linear structure that can be used to determine the direction of paleocurrent flow. Ripple marks on exposed surfaces are another means of finding the direction of ancient river flow.

FLOODPLAINS

On either side of a river channel is a floodplain (Fig. 91). It carries excess water during floods and only becomes hazardous when people build on flood-prone areas. Floodplains are the border regions of rivers and a valuable natural resource that must be properly managed to prevent flood damage. Unfortunately, the improper use of floodplains has led to destruction of property and loss of life when the inevitable flood arrives (Fig. 92 and Table 6).

Floods naturally recur. However, they are increasingly becoming man-made disasters because of the continued construction on floodplains in disregard of the flood potential. Floodplain zoning laws and flood control projects are based on short-term historical records of large floods, referred to as 50- and 100-year floods. The lack of long-term data from the geologic record often makes assessing the risk of large floods difficult. Furthermore, due to climate variability, two or more record-breaking floods can occur in consecutive years.

The purpose of floodplains is to accommodate excess water during periods of flooding. Failure to recognize this function has led to haphazard development in flood-prone areas with a consequent increase in flood dangers. Residents of North Carolina experienced the worst flood on the Tar River in 500 years. In September 1999, Hurricane Floyd turned the floodplain into a large lake and destroyed $6 billion in property. Floodplains provide level ground, fertile soils, ease of access, and water supplies. However, because of economic pressures, these areas are rapidly being developed without fully considering the flood risk.

Flood protection projects include the construction of reservoirs, whose storage capacity can absorb increased flow during floods and even out the flow rates of rivers. The dams also generate hydroelectric power, the cleanest form of energy. The reservoirs provide river navigation, irrigation, municipal water supplies, fisheries, and recreation. Without proper soil conservation measures

Figure 90 Cross bedding in outwash gravels, Rock River Valley near Janesville, Rock County, Wisconsin.

(Photo by W. C. Alden, courtesy of USGS)

in the catchment areas, however, the accumulation of silt from erosion can severely limit the life expectancy of a reservoir.

Despite flood protection programs, the average annual flood hazard is on the rise because people are moving into flood-prone areas faster than flood protection projects are being built. The increased losses therefore are not necessarily the result of larger floods but of greater encroachment onto floodplains. As the population increases, more pressure is exerted on developing flood-prone areas without proper regard for the flood potential.

About 6 percent of the land in the United States is prone to flooding (Fig. 93), and a large percentage of the nation's population and property is concentrated in these areas. More than 20,000 communities are faced with flood problems. Due to high population growth, modern floods are becoming increasingly hazardous. Besides threatening lives, floods severely damage property, destroy crops, and halt commerce.

Taking certain precautions that ultimately save lives and property can alleviate flood dangers. The factors that control flood damage include land use

Figure 91 *Parts of a floodplain that are frequently flooded include bare mud flats, Death Valley National Monument, Inyo County, California.*

(Photo by C. B. Hunt, courtesy of USGS)

on the floodplain; the depth and velocity of the floodwaters and the frequency of flooding; the rate of rise and duration of flooding; the time of year; the amount of ground saturation; and the quantity of sediment load being deposited. Effective storm forecasting also alleviates flood damage, and assists flood warning and emergency services.

The direct flood effects include injury and loss of life along with damage to buildings and other structures subjected to swift currents, debris, and sediment. In addition, sediment erosion and deposition can cause a considerable loss of soil and vegetation. Indirect flood effects include short-term pollution of rivers, the disruption of food supplies, the spread of disease, and the displacement of people who have lost their homes in the flood.

Preventing minor floods requires certain steps. Structures such as artificial levees and flood walls that serve as barriers against high water can be built. Reservoirs can be built to store excess water for later safe release (Fig. 94). Increased channel sizes can move water quickly off the land. Channels can be

Figure 92 *Severe flooding on the Pearl River, Slidell, Louisiana.*

(Courtesy of U.S. Army Corps of Engineers)

TABLE 6   CHRONOLOGY OF MAJOR U.S. FLOODS

Date	Rivers or basins	Damage in millions of dollars	Death toll
1903	Kansas, Missouri, and Mississippi	40	100
1913	Ohio	150	470
1913	Texas	10	180
1921	Arkansas River	25	120
1921	Texas	20	220
1927	Mississippi River	280	300
1935	Republican and Kansas	20	110
1936	Northeast U.S.	270	110
1937	Ohio and Mississippi	420	140
1938	New England	40	600
1943	Ohio, Mississippi, and Arkansas	170	60
1948	Columbia	100	75
1951	Kansas and Missouri	900	60
1952	Red River	200	10
1955	Northeast U.S.	700	200
1955	Pacific Coast	150	60
1957	Central U.S.	100	20
1964	Pacific Coast	400	40
1965	Mississippi, Missouri, and Red Rivers	180	20
1965	South Platte	400	20
1968	New Jersey	160	—
1969	California	400	20
1969	Midwest	150	—
1969	James	120	150
1971	New Jersey and Pennsylvania	140	—
1972	Black Hills and S. Dakota	160	240
1972	Eastern U.S.	4,000	100
1973	Mississippi	1,150	30
1975	Red River	270	—
1975	New York and Pennsylvania	300	10
1976	Big Thompson Canyon	—	140
1977	Kentucky	400	20

Date	Rivers or basins	Damage in millions of dollars	Death toll
1977	Johnstown and Pennsylvania	200	75
1978	Los Angeles	100	20
1978	Pearl River	1,000	15
1979	Texas	1,250	—
1980	Arizona and California	500	40
1980	Cowlitz and Washington	2,000	—
1982	Southern California	500	—
1982	Utah	300	—
1983	Southeast U.S.	600	20
1993	Midwest U.S.	12,000	24
1997	Red River	1,000	—
1999	Tar River	6,000	—

diverted to route floodwaters around areas requiring protection. The best method to minimize flood damage in already-developed flood-prone areas is floodplain regulation along with barriers, reservoirs, and channel improve-

Figure 93 Flood-prone areas in the United States.

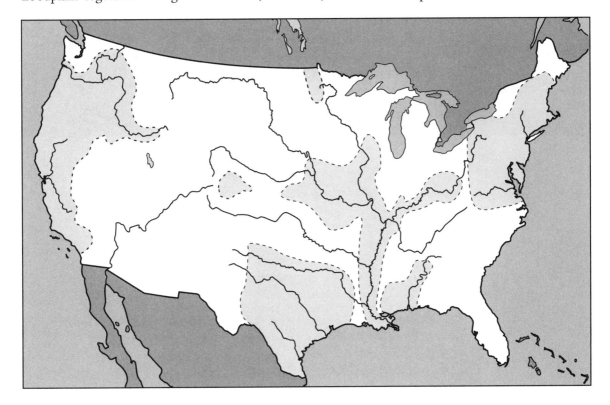

Figure 94 *A lake formed by a flood control dam, Union County, Illinois.*

(Photo by E. B. Trovillion, courtesy of USDA-Soil Conservation Service)

ments to protect lives and property better. Unfortunately, structures such as levees tend to aggravate major floods by forcing the river into a narrow channel instead of allowing the floodwaters to drain naturally onto the floodplain, which dissipates the power of the flood.

The effects of flooding are greatly reduced by developing a flood system based on providing additional natural wetlands that absorb excess floodwaters with less river engineering and levee construction. Wetlands can markedly impact the containment of most floods. However, they are less useful during great floods, especially if heavy rains have fully saturated the ground.

Communities living near rivers and streams subject to flooding should discontinue further development on floodplains that require new barriers. These barriers restrict water flow during a flood or exacerbate the effects of flooding downstream. The most practical solution is a combination of floodplain regulations and barriers that result in less physical modification of the river system. Reasonable floodplain zoning might require less flood prevention methods than the total absence of floodplain regulations.

The purpose of floodplain regulations is to obtain the most beneficial use of floodplains while minimizing flood damage and the cost of flood protection. The first step in floodplain regulations is flood hazard mapping. This provides information on land use planning in flood-prone areas. The maps delineate past floods by showing areas inundated by floodwaters and help form regulations for floodplain development.

Flood control regulations are a compromise between the indiscriminate use of floodplains, which results in the destruction of property and loss of life, and the complete abandonment of floodplains, which surrenders a valuable natural resource. Only by recognizing the dangers of flooding and making proper preparations can people safely use what nature has reserved for excess water during a flood.

FLOOD EFFECTS

Rivers are evolving entities that adapt to environmental pressures. Whenever a major flood occurs, it often reshapes the landscape through which the river flows. Floods are important for distributing soils over the land (Fig. 95). Heavily sedimented rivers also increase the severity of floods. During a flood, a river might often alter its course as it meanders along. Surface runoff cleanses the land and supplies minerals and nutrients to the ocean.

The most intense form of flooding are flash floods, which are local discharges of large volume and short duration. Torrential rains or cloudbursts during severe thunderstorms on small drainage areas generally cause them. Flash floods also result from dam breaks or by the sudden breakup of winter ice jams, releasing large volumes of flow over short periods. An unusual flash flood occurred during the May 18, 1980 eruption of Mount St. Helens. Major mudflows and extensive flooding from melted glaciers and snow on the volcano's flanks sent a torrent of mud and water laden with tree trunks into nearby rivers. Glacial outburst floods also cause severe damage downstream (Fig. 96).

In the United States, flash floods can strike in almost any part of the nation. However, they are particularly common in the mountainous areas and desert regions of the West. The Big Thompson River flood in Rocky Mountain National Park on July 31, 1976 was the worst in the region. The floods are especially hazardous in areas where the terrain is steep, surface runoff rates are high, streams flow in narrow ravines, and severe thunderstorms prevail. Flash floods during violent thunderstorms produce flooding on widely dispersed streams, resulting in high flood waves. The discharges quickly reach a maximum and diminish almost as rapidly. Floodwaters frequently contain large quantities of sediment and debris that collect as the river sweeps clean the stream channel.

Riverine floods result from heavy precipitation over large areas, by the melting of a winter's accumulation of snow, or both. They differ from flash floods in extent and duration. They occur in river systems whose tributaries drain large geographic areas and encompass many independent river basins. Floods on large river systems might continue for a few hours or for several days. They are influenced by variations in the intensity, the amount, and the distribution of precipitation. Other factors that directly affect flood runoff are the condition of the ground, the quantity of soil moisture, the vegetative cover, and the level of urbanization where impervious pavement covers the ground.

The water velocity is determined by the roughness, shape, and curving of the channel as well as by the river slope. A typical river has a slope of several hundred feet per mile near its headwaters and just a few feet per mile near

Figure 95 *A flooded farmstead in the Red River Valley, North Dakota.*

(Photo by C. Olson, courtesy of USDA)

Figure 96 *A bridge at Sheep Creek partially buried by debris from a glacier outburst flood, Copper River region, Alaska.*

(Photo by A. Post, courtesy of USGS)

its mouth. For example, the lower Mississippi River has a slope of less than half a foot per mile. A steep, rapidly flowing river such as the Colorado generally has a slope of 30 to 60 feet per mile and a slope angle of between one-third and two-thirds of a degree.

Floodwater flow is controlled by the river size and the timing of flood waves from tributaries emptying into the main channel. When a flood moves down a river system, temporary storage in the channel reduces the flood peak. As tributaries enter the main channel, the river enlarges downstream. Since tributaries are of different sizes and irregularly spaced, flood peaks reach the main channel at different times, thereby smoothing out the flow as the flood wave moves downstream and eventually empties into the sea.

Since rivers flow to the ocean and build up the coastline, the next chapter covers the coastal regions.

6

COASTAL GEOGRAPHY
LANDFORMS CREATED BY WAVE ACTION

This chapter examines coastal processes and how they affect the land. The Earth is a constantly evolving planet, with complex activities such as running water and moving waves. The shifting of sediments on the land surface and the accumulation of deposits on the seafloor continuously change the face of the Earth. The principal geologic force on the continents is erosion, whereas the primary geologic activity on the floor of the ocean is sedimentation. Rivers carry to the sea a heavy load of sediments washed off the continents, continually building up the coastal regions.

Seacoasts vary dramatically in topography, climate, and vegetation. They are places where continental and oceanic processes converge to produce a landscape that invariably changes rapidly. Sometimes human intervention is necessary to reduce the effects of naturally occurring erosional events. Regrettably, too often, these efforts are doomed to failure as waves relentlessly batter the shore.

COASTAL DEPOSITS

Rivers draining the interiors of the continents deliver to the ocean a heavy load of sediment. When reaching the coast, the river-borne sediments settle out of suspension according to grain size. The coarse-grained sediments deposit near the turbulent shore; the fine-grained sediments settle in calmer waters farther out to sea.

As the shoreline advances seaward, due to a buildup of coastal sediments or a falling sea level, finer sediments are overlain by progressively coarser ones. As the shoreline recedes landward due to a lowering of the land surface or a rising sea level, progressively finer ones overlie coarser sediments. The differing sedimentation rates as the sea transgresses and regresses result in a recurring sequence of sands, silts, and muds.

The sands comprise quartz grains about the size of beach sands. Indeed, many marine sandstone formations such as those exposed in the American West (Fig. 97) were deposited along the shores of ancient inland seas. Gravels rarely occur in the ocean and mainly move from the coast to the deep abyssal plains by submarine slides. Windblown sediments landing in the ocean slowly

Figure 97 *Cold Shivers Point, Colorado National Monument, Mesa County, Colorado.*

(Photo by J. R. Stacy, courtesy of USGS)

build deposits of abyssal red clay, whose color derived from iron oxide signifies its terrestrial origin.

The weight of the overlying sedimentary layers pressing down on the lower strata and cementing agents such as calcite and silica transform the sediments into solid rock. This creates a geologic column of alternating beds of limestones, shales, siltstones, and sandstones. Abrasion eventually grinds all rocks down to clay-sized particles, which become the most abundant sediments. The minute particles sink slowly, settling out in calm, deep waters far from shore. Compaction by the weight of the overlying strata squeezes out water between sediment grains, turning clays into mudstone if the strata is massive or into shale if the strata is fissile or thin layered.

Sediment layers vary in thickness according to the sedimentary environment at the time they were laid down. Each bedding plane marks where one type of deposit ends and another begins. Thus, thick sandstone beds might be interspersed with thin beds of shale and siltstone. This would indicate periods when coarse sediments were deposited punctuated by periods when fine sediment was laid down as the shoreline progressed and receded.

Graded bedding occurs when particles in a sedimentary bed vary from coarse at the bottom to fine at the top. This indicates rapid deposition of sediments of differing sizes by fast-flowing streams emptying into the sea. The largest particles settle out first and are covered by progressively finer material due to the difference in settling rates. Beds also grade laterally, resulting in a horizontal gradation of sediments from coarse to fine. This is called a facies change, which comes from the Latin meaning "form."

The color of sedimentary beds helps identify the type of depositional environment. Generally, sediments tinted various shades of red or brown indicate a terrestrial source, whereas green- or gray-colored sediments suggest a marine origin. The size of individual particles influences the color intensity. Darker colored sediments usually indicate finer grains.

Limestones, which once lay on the bottoms of oceans or large lakes, are among the most common rocks (Fig. 98) and make up about 10 percent of the land surface. They consist of calcium carbonate mostly derived from biologic activity, as evidenced by abundant marine-life fossils in limestone beds. Coquina is a limestone consisting almost entirely of fossils or their fragments. Some limestones chemically precipitate directly from seawater or deposit in freshwater lakes. Minor amounts precipitate in evaporite deposits formed in brine pools that periodically evaporated.

Chalk is a soft, porous, carbonate rock. One of the largest deposits is the chalk cliffs of Dorset, England, where poor consolidation of the strata results in severe erosion during coastal storms. Dolomite resembles limestone, with the calcium in the original carbonate replaced by magnesium. The chemical

Figure 98 *A limestone formation of the Bend group of the Sierra Diablo escarpment, Culberson, County, Texas.*

(Photo by P. B. King, courtesy of USGS)

reaction can cause a reduction in volume and create void spaces. The Dolomite Alps in northeast Italy are upraised blocks of this mineral deposited on the bottom of an ancient sea.

Sediments entering the ocean settle onto the continental shelf, which extends up to 100 miles or more and reaches a depth of roughly 600 feet. In most places, the continental shelf lies nearly flat, with an average slope of about 10 feet per mile, comparable to the slopes of many coastal regions. Beyond the continental shelf lies the continental slope, which plunges into the abyss to an average depth of more than two miles. It has a steep angle of several degrees, comparable to the slopes of many mountain ranges. Sediments reaching the edge of the continental shelf slide down the continental slope. Gravity slides often move huge masses of sediment that can gouge out steep submarine canyons.

On the south flank of Kilauea Volcano on the southeast coast of Hawaii, about 1,200 cubic miles of rock are moving toward the sea at geologic break-neck speeds of up to 10 inches per year. This earth movement is presently the largest on the planet and could ultimately lead to catastrophic sliding. It would be comparable to those of the past that have left massive piles of rubble on the ocean floor. Slides play an important role in building up the continental slope and the deep abyssal plains, making the seafloor one of the most geologically active places on Earth.

SEA CLIFF EROSION

Seawater lapping against the shore during severe storms provides a vivid expression of coastal erosion. The erosion of sea cliffs and dunes that mark the coastline causes the shore to retreat a considerable distance. Steep waves accompanying storms at sea seriously erode sand dunes and sea cliffs (Fig. 99). The continuous pounding of the surf also tears down most defenses erected against the raging sea.

Waves erode by impact and pressure, by abrasion, and, to a lesser extent, by solution. Therefore, wave erosion is similar to river erosion. Most beach material originates from wave erosion and river deposition. Wave impact can dislodge and transport large fragments. Waves running up on shore and back to sea move sand and pebbles back and forth, abrading sediments while simultaneously carrying them farther out to sea.

Breaking waves on a coastline develop sea cliffs by undercutting the bedrock. Coastal slides occur when wave action undercuts a sea cliff, causing it to fall into the ocean. Sea cliff retreat results from marine and nonmarine agents, including wave attack, wind-driven salt spray, and mineral solution. The

Figure 99 *The sea cliff at Moss Beach, San Mateo County, California has receded 165 feet since 1866.*

(Photo by K. R. Lajoie, courtesy of USGS)

nonmarine agents responsible for sea cliff erosion include chemical and mechanical processes, surface drainage water, and rainfall. Mechanical erosion processes rely on cycles of freezing and thawing of water in crevasses, forcing apart fractures that further weaken the rock.

Weathering agents break down rocks or cause them to shed their outer layers. Animal trails that weaken soft rock and burrows that intersect cracks in the soil also cause sea cliffs to erode. Surface water runoff and wind-driven rain further erode the sea cliff. Excessive rainfall along the coast can lubricate sediments, causing huge blocks to slide into the sea. Water running over the cliff edge and wind-driven rain produce the fluting, or grooves, often exposed on cliff faces.

Groundwater seeping from a sea cliff can form indentations on the cliff face, which undermine and weaken the overlying strata. The addition of water also increases pore pressure within sediments, reducing the shear strength that holds the rock layers together. If bedding planes, fractures, or jointing dip seaward, water moving along these areas of weakness can induce rock slides. Such slides have excavated large valleys on the windward parts of the Hawaiian Islands, where powerful springs emerge from porous lava flows.

The principal type of marine erosion is a direct wave attack at the base of a sea cliff. The waves quarry out weak beds and undercut the cliff, causing the overlying unsupported material to collapse onto the beach creating a slide (Fig. 100). Waves also work along joint or fault planes to loosen blocks of rock or soil. Furthermore, wind carrying salt spray from breaking waves drives salt water against the sea cliff. Porous sedimentary rocks absorb the salty water, which evaporates to form salt crystals whose growth weakens rocks. The surface of the cliff slowly flakes off and falls to the beach below. The material landing at the base of the sea cliff piles up into a talus cone.

Limestone cliffs erode by chemical processes that dissolve soluble minerals in the rocks. Limestone erosion commonly occurs on coral islands in the South Pacific, possibly due to rising sea levels and higher temperatures that kill the coral. This type of erosion also happens on the limestone coasts of the Mediterranean and Adriatic Seas. Seawater dissolves the lime cement in sediments, forming deep notches in the sea cliffs (Fig. 101). Chemical erosion also removes cementing agents, causing the sediment grains to separate.

BEACH EROSION

Tidal floods are overflows on coastal areas bordering the ocean or an estuary. The coastal lands, including bars, spits, and deltas, are affected by coastal currents and offer similar protection from the sea as floodplains do from rivers. Coastal flooding is primarily caused by high tides, waves from strong winds, storm surges (Fig. 102), tsunamis, or any combination of these. Tidal floods

Figure 100 *Devils slide caused by storm waves that continue to erode the base of the cliff, San Mateo County, California.*

(Photo by R. D. Brown, courtesy of USGS)

also result from high waves combined with flood runoff from heavy rains that accompany coastal storms.

Flooding can extend far along a coastline. The duration is generally short. It depends on the elevation of the tide, which usually rises and falls twice daily. If the tide is in, other factors that produce high waves can raise the maximum level of the prevailing high tide. Tidal waves generated by strong winds superimposed on regular tides produce the greatest tidal floods and the most severe beach erosion.

Much of America's once sandy beaches are sinking beneath the waves. Barrier islands and sandbars running along the Atlantic Coast and the Gulf Coast of Texas are rapidly disappearing. The shoreline on Cape Cod, Massachusetts from Nauset Spit to Highland Light has been steadily retreating. Eroding sea cliffs in California often destroy expensive homes, whose foundations are undercut by pounding waves. Most defenses erected to halt beach erosion frequently end in defeat as the sea continues to pound the shoreline.

Half the south shore of Long Island, New York is considered a high-risk zone for development. The sea is reclaiming some locations at a rate of up to six feet per year. The barrier island from Cape Henry, Virginia to Cape Hatteras, North Carolina has narrowed on both the seaward and landward sides (Fig. 103). The rest of the North Carolina Coast is rapidly retreating three to six feet annually, and much of East Texas is vanishing as well.

The strength of beach dunes or sea cliffs, the intensity and frequency of coastal storms, and the exposure of the coast influence beach erosion. Not all shoreline retreat can be blamed on sea level rise alone. It is also determined by long-term changes in the size and direction of waves striking the coast. The rate of coastal retreat varies with the geography of the shoreline and the prevailing wind and tides.

Preventing beach erosion is often thwarted because the waves constantly batter and erode defenses erected to keep out the sea. As a result, the methods developers use to stabilize the seashores are destroying the very beaches upon which they intend to build. The structures engineers build to stabilize the shoreline often aggravate beach erosion. Jetties and seawalls erected to halt the tides tend to increase erosion. Jetties cut off the natural supply of sand to beaches, and seawalls increase erosion by bouncing waves back instead of absorbing their

Figure 101 Undercut sea cliff of Pleistocene coral limestone at Port Denison, Western Australia.

(Photo by R. Revelle, courtesy of USGS)

energy. The rebounding waves carry sand out to sea, undermining the beach and destroying the shorefront property the seawall was designed to protect.

In an effort to protect houses on eroding bluffs overlooking the sea, coastal residents often construct expensive seawalls. Unfortunately, these structures tend to hasten the erosion of beach sands in front of the wall. In effect, the seawalls are saving the bluffs at the detriment of the beaches. Barriers placed at the bottom of sea cliffs might deter wave erosion but have no effect on sea spray and other erosional processes. Beaches in front of the seawalls often lose sand during certain seasons, while waves return beach sands at other times.

The disappearing beaches along the East Coast will not be resupplied with sand until the glaciers return and sea levels fall during the next ice age, when thick ice sheets lock up a significant portion of the Earth's water. Most of the sand along the coast and continental shelf as far south as Cape Hatteras originates in the north from such sources as the Hudson River and glacial deposits along Long Island and southern New England. For sand to move as far as the Carolina Coast, it must progress in stages over a period of a million years or more.

Figure 102 Storm surge penetration along Hatteras Island, Dare County, North Carolina.

(Photo by R. Dolan, courtesy of USGS)

Figure 103 *Attempts to rebuild beaches at Ocean City, Worcester County, Maryland.*

(Photo by R. Dolan, courtesy of USGS)

As the sand moves along the coast, ocean currents push it into large bays or estuaries. The embayment continues to fill with sand until sea levels drop and the accumulated sediment flushes down onto the continental shelf. The sand travels only as far as the next bay in a single glacial cycle. Therefore, most beaches will not be restocked with sand until the end of the next ice age, after sea levels return to normal.

WAVE ACTION

The breaking of a large wave on the coast is a striking example of the sizable amount of energy ocean waves generate. The intertidal zones of rocky-weather coasts receive much more energy per unit area from waves, than they do from the sun, which is about 600 watts per square yard. The waves are created by strong winds generated by distant storms blowing across large areas of the open ocean. Local storms near the coasts provide the strongest waves, especially when superimposed on rising tides.

High tides generally exceeding a dozen feet, called megatides, arise in gulfs and embayments along the coast in many parts of the world. Their height depends on the shapes of bays and estuaries, which channel the tides and increase their amplitude. Many locations with extremely high tides also experience strong tidal currents.

Most waves are generated by large storms at sea as strong winds blow across the surface of the ocean. Waves breaking on the coast dissipate energy and generate along shore currents that transport sand along the beach. High waves during coastal storms cause most beach erosion, a serious problem in areas where the shoreline is steadily receding. Sudden barometric pressure changes on large lakes or bays can cause water to slosh back and forth, producing waves called seiches. They commonly occur on Lake Michigan and, on occasions, can be quite destructive.

Hurricanes produce the most dramatic storm surges and are responsible for destroying entire beaches (Fig. 104). As the wave approaches shore, it touches bottom and slows. The shoaling of the wave distorts its shape, making it break upon the beach. The breaking wave dissipates its energy along the coast and erodes the shoreline.

Wave energy reflecting off steep beaches or seawalls forms sandbars. When waves approach the shore at an angle to the beach, the wave crests bend by refraction. As waves pass the end of a point of land or the tip of a breakwater, a circular wave pattern generates behind the breakwater. When the refracted waves intersect other incoming waves, they increase the wave height.

Swells reaching a coast produce various types of breakers, depending on the wave steepness and bottom slope conditions near the beach. If the bottom slope is relatively flat, the wave forms a *spilling breaker,* the most common type. It is an oversteepened wave that starts to break at the crest and continues breaking as it heads toward the beach.

If the bottom slope increases to about 10 degrees, the wave forms a *plunging breaker* (Fig. 105) and the crest curls over, creating a tube of water. As the wave breaks, the tube moves toward the bottom and stirs up sediments.

Figure 104 *Storm surge damage from Hurricane Frederic on September 12, 1979 at Gulf Shores, Baldwin County, Alabama.*

(Courtesy of USGS)

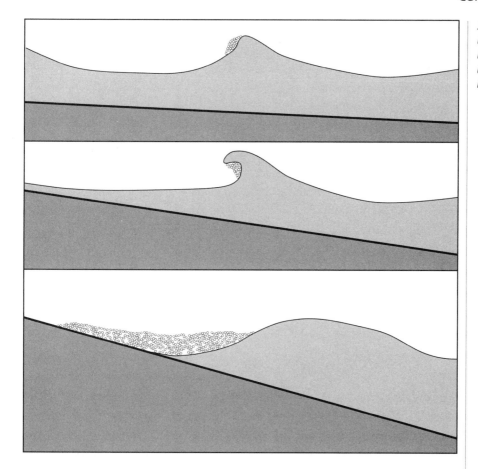

Figure 105 *Types of breakers. Top, spilling breaker; middle, plunging breaker; bottom, surging breaker.*

Plunging waves are the most dramatic breakers and do the most beach damage because the energy concentrates at the point where the wave breaks.

If the bottom slope steepens to about 15 degrees, the wave forms a collapsing breaker. The breaker is confined to the lower half of the wave. However, as the wave moves toward the coast, most of it reflects off the beach. On a steep bottom where the slope is greater than 15 degrees, a *surging breaker* develops. The wave does not break but surges up the beach face and reflects off the coast, generating standing waves near the shore. Standing waves are important for the development of offshore structures such as bars, sand spits, beach cusps, and rip tides.

The world's most powerful waves, called seismic sea waves or tsunamis, are generated by undersea earthquakes, volcanic eruptions, or coastal landslides. The vertical displacement of the ocean floor during an earthquake causes the most destructive tsunamis, whose wave energy is determined by the intensity of the quake. The powerful 1960 earthquake off the coast of Chile elevated a California-sized chunk of land about 30 feet and sent deadly tsunamis racing across the Pacific Ocean. Between 1992 and 1996, 17 tsunami attacks around the Pacific killed some 1,700 people.

Volcanic eruptions at sea also set up tsunamis that pounce on nearby coasts. The powerful waves transmit the volcano's energy to areas outside the direct reach of the volcano itself. Tsunamis are also produced by large pyroclastic flows into the ocean or by landslides triggered by erupting volcanoes. Undersea flow failures also generate large tsunamis that overrun parts of the coast.

A tsunami extends thousands of feet to the ocean floor and speeds across the sea at 500 miles per hour or more. As the tsunami wave touches bottom in a harbor or a narrow inlet, its speed diminishes abruptly to about 100 miles per hour. This sudden breaking action piles up water, magnifying the wave height to several tens of feet. The destructive force of the wave is immense. The damage it causes as it crashes to shore can be devastating (Fig. 106).

Figure 106 *Tsunami damage at Lebak in Moro Gulf, Island of Mindanao, Philippines on August 16, 1976.*

(Photo by R. E. Wallace, courtesy of USGS)

COASTAL SUBSIDENCE

Subsidence is the downward settling of earth materials with little horizontal movement. Coastal areas often subside during large earthquakes when one block of crust drops below another. During an earthquake, vegetated lowlands elevated to avoid inundation by the sea submerge regularly and become barren tidal mud flats (Fig. 107). Between quakes, sediments fill the tidal flats, raising them to the level where vegetation can grow once again. Therefore, repeated earthquakes produce alternating layers of lowland soil and tidal flat mud.

In the United States, earthquake subsidence occurs mainly in California, Alaska, and Hawaii, which are highly prone to large earthquakes. The subsidence is caused by vertical displacements along faults that can affect broad areas. During the 1964 Alaskan earthquake, over 70,000 square miles of land tilted downward, causing extensive flooding in coastal areas of southern Alaska. The earthquake produced submarine flow failures that destroyed many seaport facilities. The flow failures also generated large tsunamis that overran coastal areas, causing additional damage and casualties. Coastal regions in Japan are particularly susceptible to subsidence. Had the January 17, 1995 Kobe earthquake of 7.2 magnitude struck Tokyo instead, more than half the city would have sunk beneath the waves.

Subsidence also occurs when fluids pumped from subterranean sediment result in compaction. Some of the most dramatic examples of subsidence occur along the seacoasts (Fig. 108). Coastal cities subside due to a combination of rising sea levels and withdrawal of groundwater, causing the aquifer to compact. Subsidence in some coastal towns has increased susceptibility to flooding during severe coastal storms. Venice, Italy has been fighting a losing battle against the sea as the city continues to sink from overuse of groundwater while the ocean rises.

COASTAL INUNDATION

Civilizations have had to cope changing sea levels for centuries (Table 7). The melting of the polar ice caps during a sustained warmer climate could substantially raise sea levels and drown coastal regions. At the present rate of melting, the sea could rise a foot or more by the middle of the 21st century, comparable to the melting rate of the continental glaciers at the end of the last ice age. Consequently, beaches and barrier islands inevitably disappear as shorelines move inland (Fig. 109).

The global sea level has risen half a foot over the past century mainly by the melting of the Antarctic and Greenland ice caps (Fig. 110) along with alpine glaciers. Some areas, such as the European Alps, might have lost more than half their cover of ice. Moreover, the rate of loss appears to be accelerating. Tropical glaciers have receded at a rate of 150 feet per year over the last two decades.

In areas such as Louisiana, the level of the sea has risen upward of three feet per century. The thermal expansion of the ocean has also raised the sea level about two inches. Surface waters off the California coast have warmed

Figure 107 *Secondary cracks in the tidal flat at the head of Bolinas Lagoon, Marin County, California from the 1906 California earthquake.*

(Photo by G. K. Gilbert, courtesy of USGS)

nearly one degree Celsius over the past half-century, causing the water to expand and raise the sea about 1.5 inches.

Higher sea levels are also caused, in part, by sinking coastal lands due to the increased weight of seawater pressing down on the continental shelf. In addition, sea level measurements are affected by the rising and sinking of the land surface due to plate tectonics and the rebounding of the continents after glacial melting at the end of the last ice age.

If all the polar ice melted, the additional seawater would move the shoreline as far as 70 miles inland in most places and much farther at low-lying areas such as river deltas. The inundation would radically alter the shapes of the continents. All of Florida along with south Georgia and the eastern Carolinas would vanish. Mississippi, Louisiana, eastern Texas, and major parts of Alabama and Arkansas would virtually disappear. Much of the isthmus separating North and South America would sink out of sight.

Figure 108 *The subsidence of the coast at Halape from the November 29, 1975 Kalapana earthquake, Hawaii County, Hawaii.*

(Photo by R. I. Trilling, courtesy of USGS)

TABLE 7   MAJOR CHANGES IN SEA LEVEL

Date	Sea level	Historical event
2200 B.C.	Low	
1600 B.C.	High	Coastal forest in Britain inundated by the sea.
1400 B.C.	Low	
1200 B.C.	High	Egyptian ruler Ramses II builds first Suez canal.
500 B.C.	Low	Many Greek and Phoenician ports built around this time are now under water.
200 B.C.	Normal	
A.D. 100	High	Port constructed well inland of present-day Haifa, Israel.
A.D. 200	Normal	
A.D. 400	High	
A.D. 600	Low	Port of Ravenna, Italy becomes landlocked. Venice is built and is presently being inundated by the Adriatic Sea.
A.D. 800	High	
A.D. 1200	Low	Europeans exploit low-lying salt marshes.
A.D. 1400	High	Extensive flooding in low countries along the North Sea. The Dutch begin building dikes.

Figure 109 Beach wave erosion at Grand Isle, Jefferson Parish, Louisiana on August 30, 1985 from Hurricanes Danny and Elana.

(Courtesy of U.S. Army Corps of Engineers)

Every foot of sea level rise would inundate up to 1,000 feet of seashore, depending on the slope of the coastline. A three-foot rise in sea level could flood about 7,000 square miles of coastal land in the United States, including most of the Mississippi Delta, as seas reach the outskirts of New Orleans. Half the scattered islands of the Republic of Maldives southwest of India would be lost. Much of Bangladesh would drown as well.

If the ocean continues to rise, the Dutch, who worked so hard building dikes to reclaim land from the sea, would find a major portion of their country underwater. Many islands would disappear or become mere skeletons, with only their mountainous backbones protruding above the sea. Because most major cities of the world are either located on seacoasts or along inland waterways, they would be inundated by the ocean, with only the tallest skyscrapers poking above the waterline.

The receding shore would result in the loss of large tracks of coastal land along with shallow barrier islands. Estuaries, where marine species hatch their young, would be destroyed. The rising waters would inundate low-lying fertile deltas that feed much of the world's population. Coastal cities, where half

Figure 110 U.S. Air Force radar station at Thule, Greenland.

(Courtesy of U.S. Air Force)

137

the world's human population reside, would have to move farther inland or build seawalls to keep out the rising waters.

BARRIER REEFS

Coral reefs are among the oldest ecosystems on Earth and are important land builders, forming chains of islands and altering continental shorelines. An estimated 270,000 square miles of coral reefs exist in the world's oceans. Over geologic time, corals and other organisms living on the reefs have built massive formations of limestone. A typical reef consists of fine, sandy detritus stabilized by plants and animals anchored to the surface. The coral's ability to build wave-resistant structures encourages tropical plant and animal communities to thrive on the reefs, which are thought to house one in every four marine species.

The reefs are limited to clear, warm, sunlit tropical waters in the Indo-Pacific and the western Atlantic (Fig. 111). Hundreds of atolls, comprising rings of coral islands that enclose a central lagoon, dot the Pacific Ocean. They consist of reefs several thousand feet across. Many of these formed on ancient volcanic cones that have dropped beneath the waves. The rate of coral growth evenly matches the rate of subsidence.

The major structural feature of the living reef is the coral rampart, which reaches almost to the water's surface. It consists of large, rounded coral heads and a variety of branching corals. Hundreds of species of encrusting organisms such as barnacles thrive on the coral reef. Smaller, more fragile corals and large communities of green and red calcareous algae live on the coral framework (Fig. 112).

Figure 111 *The extent of coral reefs is shown in the stippled area.*

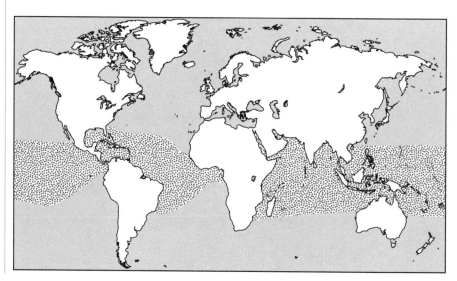

Figure 112 *The coral-algal zone on a reef fringing Agana Bay on the island of Guam.*

(Photo by J. I. Tracy, courtesy of USGS)

The fore reef is seaward of the reef crest, where coral blankets nearly the entire seafloor. In deeper waters, many corals grow in flat, thin sheets to maximize their light-gathering area. In other parts of the reef, the corals form large buttresses separated by narrow, sandy channels composed of calcareous debris from dead corals, calcareous algae, and other organisms living on the coral.

The channels resemble narrow, winding canyons with vertical walls of solid coral. They dissipate wave energy and allow the free flow of sediments. This prevents the coral from choking on the debris. Below the fore reef is a coral terrace, followed by a sandy slope with isolated coral pinnacles, then another terrace, and finally a nearly vertical drop into the dark abyss.

Fringing reefs grow in shallow seas and hug the coastline or are separated from the shore by a narrow stretch of water. Barrier reefs also parallel the coast but lie farther out to sea. They are much larger and extend for longer distances. The best example is the Great Barrier Reef, a chain of more than 2,500 coral reefs and islands off the northeastern coast of Australia. It forms an undersea embankment more than 1,200 miles long, up to 90 miles wide, and as much as 400 feet high. The reef is the largest feature built by living organisms and one of the great wonders of the world.

The global sea level rose and fell more than 30 times between 6 million and 2 million years ago. The changes in sea level during the last million years have produced terraces of coral growth running up a continent or an island. The drowned coral represent periods of extensive glaciation, when sea levels dropped as much as 400 feet. In Jamaica, nearly 30 feet of reef have built up since the present sea level stabilized some 5,000 years ago when the glaciers from the last ice age retreated to the poles.

The next chapter will turn the discussion to deserts and their unique landforms.

7

DESERT FEATURES

LANDFORMS CREATED BY BLOWING SAND

This chapter examines desert landscapes and discusses how desertification and dust storms shape the land. Deserts are more than just barren landscapes mostly devoid of vegetation and animal life. They are among the most dynamic landforms, constantly changing by roving sands. Sometimes, sand dunes cover over human settlements and other man-made features, often causing considerable damage. Sandstorms are particularly hazardous. Thousands of tons of sediment clogs the skies and lands in places where it is not wanted.

About one-third of the Earth's landmass is covered by deserts (Fig. 113 and Table 8). They are the hottest and driest regions and among the most barren environments. The world's desert wastelands receive only minor amounts of precipitation during certain seasons, whereas some areas have been essentially rainless for years. Only the hardiest of plant and animal species, some with very unusual adaptations, can tolerate these arid conditions.

Often, when the rains arrive, heavy downpours cause severe flash floods that sweep away massive quantities of sediment and debris. Gigantic dust storms and sandstorms prevalent in desert regions also play a major role in shaping the arid landscape. Roving sand dunes driven across the desert by strong winds engulf everything in their paths.

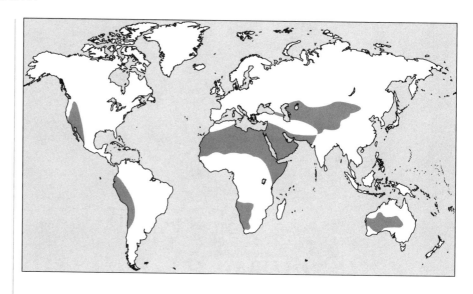

Figure 113 *Location of worldwide deserts.*

ARID REGIONS

About 30 percent of the land surface, or roughly 20 million square miles, is wilderness undisturbed by people and with few signs of human habitation. Several broad belts of wilderness wind around the globe, including a band of deserts running southwest from far-eastern Asia through Tibet, Afghanistan,

TABLE 8   MAJOR DESERTS

Desert	Location	Type	Area (square miles x 1,000)
Sahara	North Africa	Tropical	3,500
Australian	Western/Interior	Tropical	1,300
Arabian	Arabian Peninsula	Tropical	1,000
Turkestan	Central Asia	Continental	750
North American	Southwest U.S./North Mexico	Continental	500
Patagonian	Argentina	Continental	260
Thar	India/Pakastan	Tropical	230
Kalahari	Southwest Africa	Littoral	220
Gobi	Mongolia/China	Continental	200
Takla Makan	Sinkiang, China	Continental	200
Iranian	Iran/Afganistan	Tropical	150
Atacama	Peru/Chile	Littoral	140

and Saudi Arabia into Africa. The forbidding Sahara Desert in northern Africa and the great central desert of Australia are among the least-densely populated regions in the world. The drylands bordering the deserts cover a quarter of the landmass and support about 15 percent of the human population.

Around 4,000 years ago, temperatures dropped significantly following an unusually warm period known as the Climatic Optimum. The global climate grew more arid, spawning the deserts of today. Deserts typically receive less than 10 inches of rainfall annually. Evaporation generally exceeds precipitation throughout most of the year (Fig. 114). Human activities and natural processes cause additional land to become desertified, amounting to about 15,000 square miles a year. This is an area slightly less than the size of California's Mojave Desert (Fig. 115) at the south end of the Sierra Nevada Range.

Most of the world's deserts lie in the subtropics in a broad band between 15 and 40 degrees latitude on either side of the equator. High precipitation levels in the tropics leave little moisture for the subtropics, where the dry air cools and sinks. This produces zones of semipermanent high pressure called blocking highs because they tend to block advancing weather systems from entering the region. The resulting compression heats the dry desert air, while the high pressure produces mostly clear skies and calm winds.

Mountains also block weather systems by forcing rain clouds to rise and precipitate on the windward side of the range (Fig. 116). The lack of precipitation on the lee side or opposite end of the mountains results in a rain shadow zone that is rain deficient, creating deserts such as those in the southwestern United States. Moist winds from the Pacific cool and precipitate as they rise over the Sierra Nevada and other mountain ranges in California, leaving regions to the east parched and dry.

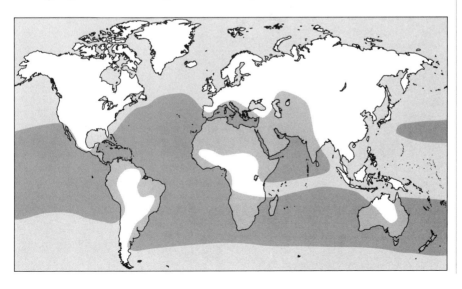

Figure 114 *The global precipitation-evaporation balance. In the darkened areas, evaporation exceeds precipitation.*

Because of their latitude and climate patterns, much of the world's desert wasteland receives only minor amounts of rainfall during certain seasons. In fact, some regions, such as Egypt's Western Desert, have gone years without significant rainfall. Often, during the rainy season, heavy downpours cause severe flash floods that scour the desert landscape. Water levels in dry washes or wadis rise rapidly and fall almost as fast, as the flood wave flows through the desert. Eventually, the floodwaters empty into shallow lakes and evaporate or soak into the dry, parched ground. Sometimes mudflows raft boulders out of the mountains onto the desert floor (Fig. 117).

Light-colored deserts have a high albedo (Table 9) and reflect most sunlight striking the surface. Desert sands absorb heat during the day, when the surface scorches at temperatures often exceeding 65 degrees Celsius (150 degrees Fahrenheit). However, because the skies are usually clear at night, the

Figure 115 *Eastern Mojave Desert, San Bernardino County, California.*

(Photo by R. E. Wallace, courtesy of USGS)

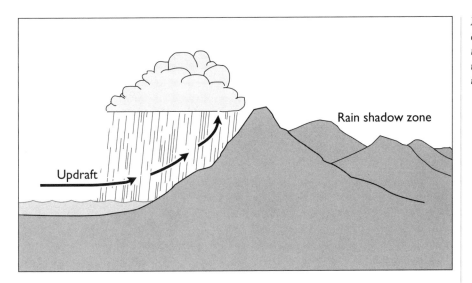

Rain shadow zone

Updraft

thermal energy trapped in the sand quickly escapes due to its low heat capacity. Therefore, desert regions are among the coldest places. Even summertime temperatures at high elevations can drop significantly at night. Consequently, deserts have the greatest temperature extremes of any environment.

DESERT GEOGRAPHY

A few times a year during heavy rains, a dry lake bed called the Racetrack Playa at the north end of Death Valley turns into a shallow lake. The area is known for its mysterious roving rocks, which leave tracks in the lake bed that have puzzled geologists for half a century. Originally, high winds whistling off the nearby mountains were thought to be powerful enough to push the boulders through the muddy lake bed after a soaking rain. Yet the largest of the boulders measured 2 feet across and weighed some 700 pounds, much too heavy for the wind to push around. If, however, a thin layer of ice formed on the lake bed after a winter rain, it could lift the rocks slightly, reducing the frictional contact with the mud. With the aid of a strong wind, the inch-thick sheet of ice would skim across the lake bed, with embedded rocks etching patterns in the mud.

Other examples of patterned ground include polygonal shapes created in Death Valley's desert muds (Fig. 118). The mud cracks formed when mud contracted as it rapidly dried in the hot desert sun. Sorted circles were also created by the increased wearing down of coarse grains in isolated cracks in bedrock. Even earthquake vibrations are believed to cause the sorting of some sediment.

The erosion of desert mountain ranges relies on heavy downpours along small drainage areas. Sediment fans consisting of sands and gravels develop at the mountain front. When the formally steep mountain front retreats, it leaves a smooth surface in the bedrock called a pediment, which generally has a concave-upward slope of up to seven degrees, depending on the sediment size and the amount of runoff. Streams issuing from the mountains change course back and forth across the pediment in a manner similar to the formation of alluvial fans. Eventually, the mountains erode down to the level of the plain, leaving the pediments speckled with remnants of the range.

The development of drainage patterns in desert lands is well demonstrated in the Basin and Range Province of the American Southwest. The region contains several mountain ranges formed by relatively recent faulting. The basins between ranges are low-lying areas that often contain lakes during wetter climates. Lake-deposited sediments commonly occur, in most deserts

TABLE 9   ALBEDO OF VARIOUS SURFACES

Surface	Percent reflected
Clouds, stratus	
< 500 feet thick	25–63
500–1,000 feet thick	45–75
1,000–2,000 feet thick	59–84
Average all types and thicknesses	50–55
Snow, freshly fallen	80–90
Snow, old	45–70
White sand	30–60
Light soil (or desert)	25–30
Concrete	17–27
Plowed field, moist	14–17
Crops, green	5–25
Meadows, green	5–10
Forests, green	5–10
Dark soil	5–15
Road, blacktop	5–10
Water, depending upon sun angle	5–60

and dry lake beds called playas cover the surface (Fig. 119). The bodies of water are called alkali lakes because of their high concentrations of salt and other soluble minerals. When the lakes evaporate, they become alkali flats such as Utah's Bonneville Salt Flats.

At 1,300 feet below sea level, the Dead Sea in the Syrian Desert on the border between Israel and Jordan is the lowest place on land. It occupies the Jordan Rift Zone, a deep gash in the crust where the land is being pulled apart. It is also one of the world's deepest lakes, some 1,000 feet in depth. For thousands of years, rivers ladened with salts leached from the rocks flowing south through the Jordan Rift Valley terminated in the Dead Sea. With no outlet, the inflowing water evaporates into the dry desert air, which concentrates the salts, making the Dead Sea the world's saltiest lake. Its average salinity is eight times higher than the ocean's.

The Sahara Desert (Fig. 120), at 3.5 million square miles—about the size of the United States, is the largest arid region on Earth. Deep beneath its sands lies a vast network of ancient river valleys and smaller stream channels that

Figure 118 *Pattern of scalloped cracks near Stovepipe Wells, Death Valley National Monument, California.*

(Courtesy of National Park Service)

wind across the bedrock along with gravel terraces, desert basins, and other geologic structures. A search through the sands uncovered one of the last great river systems in the world as wide as Egypt's Nile Valley. Channels hundreds of thousands of years old wind through valleys millions of years old. The discovery suggests that buried riverbeds exist in other deserts as well. They might be good sources for groundwater, a commodity desperately needed in these parched regions.

The buried valleys lying under the sand might once have been migration routes for early humans leaving Africa for Europe and Asia. Scattered in the sands of the Sahara were Stone Age artifacts, suggesting humans and life-sustaining water sources once existed in an area that is now utterly uninhabitable. Dozens of human artifacts, including stone axes up to a quarter-million years old, appear to mark ancient campsites, where people known as *Homo erectus* lived and made stone tools.

While exploring for petroleum in the Sahara, geologists stumbled upon a series of giant grooves apparently cut into the underlying strata by glaciers during an ancient glaciation. Rocks embedded at the base of the glacial ice scoured the landscape as the massive ice sheets moved back and forth. Evidence suggesting that thick sheets of ice once blanketed the region include erratic boulders dumped in heaps by the glaciers and eskers, which are long, sinuous sand deposits from glacial outwash streams.

The coldest regions and possibly the most impoverished deserts on Earth are the dry valleys running between McMurdo Sound and the Transantarctic Range in Antarctica (Fig. 121). Because the mountains protect them against snowstorms, the dry valleys receive less than four inches of snow annually. Most of it blows away by hurricane force winds reaching 200 miles per hour and more. Some areas have not received any form of precipitation for as long as a million years.

Figure 119 *Silver Lake Playa, San Bernardino County, California.*

(Photo by D. G. Thompson, courtesy of USGS)

Figure 120 *Linear dunes crested with barchanoid ridges in the northwest Sahara Desert, Algeria, northern Africa.*

(Photo by E. D. McKee, courtesy of USGS and NASA)

DESERTIFICATION

Throughout the world, nearly twice the area of the United States has turned to desert since the beginning of agriculture. This occurred mostly from the abuse of the land. In North America alone, an estimated 1.1 billion acres have been desertified. Much of the American West 150 years ago was almost uninterrupted grassland that has since become desert.

In the next two decades, perhaps as much as half a million square miles (about the size of Alaska) of the world's valuable agricultural land could be destroyed by desertification. Tropical rain forests have dwindled over the last few decades. Man-made deserts have replaced once vast expanses of trees. Severe erosion from large-scale deforestation clogs rivers with sediment, disrupting the lives of downstream residents. Africa has the worst soil erosion problem in the world. Its rivers are the most heavily polluted with sediment, while other streams have completely dried out.

Yearly, the deserts claim additional land (Fig. 122). The process of desertification, which mainly results from human activity and climate, severely degrades the environment by removing precious topsoil. Soil erosion takes out of production millions of acres of once fertile cropland and pasture every year. Worldwide, perhaps one-third or more of the productive land has been rendered useless by erosion and desertification. After the topsoil disappears, only the coarse sands of the infertile subsoil remain, creating desert conditions.

Desertification is a global problem, but it is most severe in Central Africa. The sands of the Sahara Desert march back and forth across the Sahel region to the south, once a stretch of forests and grasslands. Desertification is also self-perpetuating because the light-colored sands reflect sunlight very well. This creates high-pressure zones that block weather systems and reduce

Figure 121 Mountains *and dry valleys in* Antarctica.

(Photo by F. R. Bair, courtesy of U.S. Navy)

Figure 122 *A terrace structure in the leeward side of sand dunes, Point Ano Nuevo, San Mateo County, California.*

(Photo by R. Arnold, courtesy of USGS)

rainfall. The lesser rainfall denudes or strips the land of its vegetation, enabling deserts to creep across previously fertile areas.

The land is subjected to flash floods, higher erosion and evaporation rates, and dust storms that transport the soil out of the region. Desertification is also exacerbated by the lack of vegetation because roots hold the soil in place. Perhaps in a million years or more, when climatic conditions allow, the deserts will revert back to their original environments.

Droughts occur when precipitation patterns shift around the world. If global temperatures continue on their upward trend, the central regions of continents that normally experience occasional droughts could become permanently dry wastelands. Strong winds would create horrific dust storms and inflict severe erosional problems.

The tendency of the wind to erode the soil is often aggravated by improper agricultural practices. In the United States, wind erodes about 20 million tons of soil per year. Wind erosion removes from production an estimated 1.2 million acres of farmland in Russia annually, exacerbating the nation's inability to feed itself. The primary method of controlling wind erosion is by maintaining a surface cover of vegetation. However, if rainfall is inadequate, these measures might be futile. The soil will simply blow away.

DUST STORMS

Dust storms are awesome meteorological events that can directly threaten life. People and animals have been known to suffocate to death during severe dust storms. Another direct threat posed by dust storms is soil erosion. A dust bowl is a region that suffers from prolonged droughts and dust storms. The 1930s Dust Bowl in the American Midwest was the nation's worst ecologic disaster (Fig. 123). Topsoil was airlifted out of the Great Plains and deposited elsewhere, often burying areas under thick layers of sediment. Massive dust storms raced across the prairie, transporting over 150,000 tons of sediment per square mile. The Dust Bowl years exacted a heavy toll on U.S. agricultural resources, and a repeat of those conditions today could be devastating.

Vast dust storms occur when an enormous airstream moves across deserts, particularly those in Africa. Similar dust storms also arise in Arabia, central Asia, central China, and the deserts of Australia and South America. In Africa, giant dust bands 1,500 miles long and 400 miles wide often traverse the region and are driven by strong cold fronts.

Some large storm systems can carry dust out of Africa to South America, where about 13 million tons land in the Amazon Basin annually. The dust over African deserts rises to high altitudes, where westward-flowing air currents transport it across the Atlantic Ocean. Fast-moving storm systems in the

Figure 123 *The 1936 Dust Bowl days in Cimarron County, Oklahoma.*

(Photo by A. Rothstein, courtesy of USDA–Soil Conservation Service)

Amazon Rain Forest pull in the dust, which contains nutrients that enrich the soil. So much African dust blows across the Atlantic during summer storms that areas along the East Coast of the United States actually violate clean-air standards. The dust has a distinctive red-brown color. When added to other air pollutants, it causes a persistent haze, especially in summer.

Dust storms often form a solid wall of sediment blowing at speeds of up to 60 miles per hour or more. Like avalanches, dust storms build up as they travel. The rising dust clouds can extend several thousand feet high and stretch hundreds of miles. As the winds scour the land, several inches of soil is carried to other areas, causing major problems downwind. Once airborne, dust particles travel quite far, even encircling the globe.

Severe dust storms form over deserts during large thunderstorms (Fig. 124), such as those in northern Africa. They accompany the rainy season, removing a remarkable amount of sediment. A typical dust storm of 300 to 400 miles in diameter lifts over 100 million tons of sediment. During the height of the season between May and October, several feet of sand piles up against obstructions exposed to the full fury of the storm.

Many severe dust storms occur in the American Southwest, where Phoenix, Arizona averages about a dozen annually. As with those in Africa,

Figure 124 *The structure of a dust storm.*

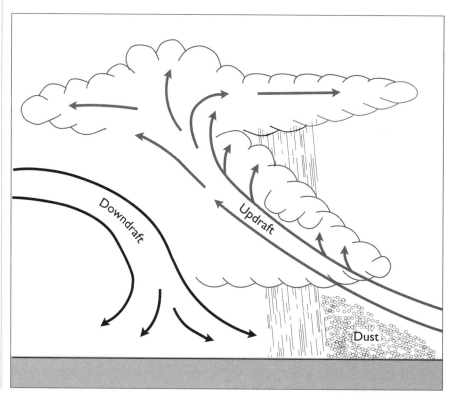

American dust storms most frequently occur during the rainy season, normally July and August. Surges of moist, tropical air from the Pacific flow out of the Gulf of California into Arizona and generate long, arching squall lines, with dust storms fanning out in front.

Individual outflows often merge to form a solid wall of sand and dust stretching hundreds of miles. The sediment rises 10,000 to 15,000 feet above the ground, traveling at an average speed of 30 miles per hour and gusts of 60 miles per hour or more are possible. Dust storms also generate small, short-lived, and intense whirlwinds or dust devils within the storms themselves or a short distance out in front.

While a dust storm is in progress, average visibility falls to about a quarter mile and lowers to near zero in very intense storms. After the storm passes, the skies begin to clear, and visibility returns to normal after an hour or more. If the parent thunderstorm arrives behind the dust storm, its precipitation quickly clears the air. Often, though, the trailing thunderstorm fails to show or the precipitation evaporates before reaching the ground. This phenomenon is known as virga, and the lack of rainfall causes the sediment to remain suspended for hours or even days.

WIND EROSION

Only in desert regions is wind an active agent of erosion, transportation, and deposition. The deserts host some of the strongest winds due to rapid heating and cooling of the land surface. The winds generate sandstorms and dust storms, which work together to cause wind erosion. Wind erosion develops mainly by the removal of large amounts of sediment during windstorms, forming a deflation basin. In some areas, the wind scours out concavely shaped hollows called blowouts (Fig. 125).

Wind erosion causes deflation and abrasion. Deflation is the removal of sand and dust particles by the wind, which often excavates hollowed-out areas. It usually occurs in arid regions and unvegetated areas such as deserts and dry lake beds. As smaller soil particles blow away during dust storms, the ground coarsens over time. The remaining sand tends to roll, creep, or bounce with the wind until it meets an obstacle, whereupon it settles and builds into a dune. Abrasion is similar to sandblasting by wind-driven sand grains and can cause erosion near the base of a cliff. When acting on boulders or pebbles, abrasion pits, etches, grooves, and scours exposed rock surfaces. Abrasion also produces some unusually shaped rocks called ventifacts, which often have several flat, polished surfaces depending on the wind direction or the movement of the rock.

Sandblasting erodes the surface of boulders. Desert varnish, which is composed of iron and magnesium oxides exuded from the rock, colors them

Figure 125 *A wind blowout in Fremont County, Idaho in August 1921.*

(Photo by H. T. Stearns, courtesy of USGS)

dark brown or black. Maximum erosion effects occur during strong sandstorms, with sediment grains generally rising less than two feet above the ground. These abrasive effects occur most commonly on fence posts and power poles.

Sands behave in mysterious ways, acting partly as solids and partly as liquids. Sand grains march across the desert floor under the influence of strong winds by a process called saltation (Fig. 126). The grains of sand become airborne for an instant, rising a foot or more above the ground. When landing, they dislodge additional sand grains, which repeat the process. The rest of the moving sand travels forward along the surface by rolling and sliding. The sediment grains of desert deposits are often frosted due to abrasion from the constant motion of the sand.

Often wind erosion removes the fine material from the surface, leaving a layer of pebbles to prevent further erosion. Over a period of thousands of years, deserts develop a protective shield of pebbles coated with desert varnish. The pebbles vary in size from a pea to a walnut and are too heavy for the strongest desert winds to pick up. The desert shield thus helps hold down sand grains and create a stable terrain. Any disturbance on the surface can spawn a new generation of roving sand dunes.

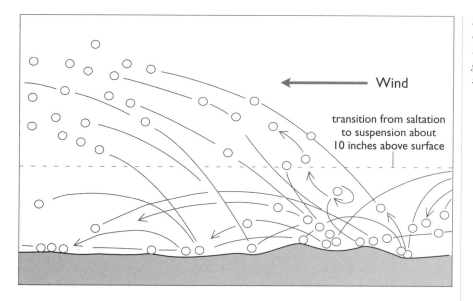

Wind

transition from saltation
to suspension about
10 inches above surface

SAND DUNES

About 10 percent of the world's arid regions contain sand dunes (Fig. 127), that are driven across the desert by powerful wind currents. The dunes move across the desert floor in response to the wind as sand grains in motion dislodge one another and become airborne for a moment. The direction, strength,

Figure 127 *Red sand dunes and tan playa sediments along the interior margin of the Namib Desert, Southwest Africa.*

(Photo by E. D. McKee, courtesy of USGS)

and variability of the wind, the soil's moisture content, the vegetative cover, and the underlying topography determine the size and shape of sand dunes. They also determine the quantity of movable soil exposed to the wind.

As sand dunes march across the desert floor, they engulf everything in their paths. This causes major problems in the construction and maintenance of highways and railroads that cross sandy areas of deserts. Sand dune migration near desert oases creates another serious problem, especially when encroaching on villages. Damages to structures from sand dunes can be reduced by erecting windbreaks and by funneling sand out of the way. Without such measures, disruption of roads, airports, agricultural settlements, and towns could pose many difficulties for desert regions.

Sand dunes generally acquire three basic shapes determined by the topography of the land and the patterns of wind flow. Linear dunes (see Fig. 120 on page 150) align in roughly the direction of strong, steady, prevailing winds. Linear dunes are significantly longer than they are wide. They parallel each other, sometimes producing a wavy pattern. When the wind blows over the dunes' peaks, part of the air flow shears off and turns sideways. The air current scoops up sand and deposits it along the length of the dune, which maintains and lengthens it. The surface area covered by dunes is about equal to the area between dunes. Both sides of the dune are likely steep enough to cause avalanches.

Crescent dunes, also called barchans, are symmetrically shaped with horns pointing downwind. They travel across the desert at speeds of up to 50 feet a year. Parabolic dunes form in areas where sparse vegetation anchors the side arms while the center blows outward and moves sand in the middle forward. Star or radial dunes (Fig. 128) form by shifting winds that pile sand into central points that can rise 1,500 feet and more. Several arms project outward, resembling giant pinwheels. Sand also accumulates in flat sheets or forms stringers downwind that do not exhibit any appreciable relief in sand seas.

One curious feature of sand dunes that march across the desert floor is an unexplained phenomenon known as booming sands. When sand slides down the lee side of a dune, it sometimes emits a loud rumbling boom, similar to the sound of a jet aircraft flying overhead. Simply walking along the dune ridges can trigger the booms. The low-frequency sound appears to originate from a cyclic event occurring at an equally low frequency. However, normal landsliding involves a mass of randomly moving sand grains that collide with a frequency much too high to produce such a peculiar noise.

LOESS DEPOSITS

Thick deposits of fine, windblown sediment called loess carpet vast areas downwind from glaciated regions. They appear to have formed when glacial

Figure 128 *Star dunes southeast of Zalim, Saudi Arabia.*

(Photo by E. D. McKee, courtesy of USGS)

retreat left large, unvegetated areas subjected to wind erosion. During glaciation, many regions not covered by glaciers dried out due to the lack of rainfall, resulting in widespread desertification. The lower precipitation levels caused by the colder climate expanded the area of arid zones and increased the size of the world's deserts. Strong winds blowing across desert regions produced gigantic dust storms that raised massive clouds of dust. The dust clogging the atmosphere also shaded the Earth and helped maintain a cool climate in an effective feedback mechanism that sustained the Ice Age.

In dry regions where dust storms prevailed, the wind transported large quantities of loose sediment. Most windblown sediments accumulated into thick deposits of loess (Fig. 129), which is a fine-grained, loosely consolidated, sheetlike deposit that often shows thin, uniform bedding on outcrop. Secondary loess deposits were transported and reworked over a short distance by water or were intensely weathered in place.

Loess deposits cover thousands of square miles and were laid down during the Pleistocene Ice Ages, when continental glaciers swept out of the Arctic regions and buried much of the northern lands. The loess was derived primarily from outwash in the vicinity of major streams that carried glacial meltwater from the front of the glacier. The retreating ice left large, unvegetated areas adjacent to rivers susceptible to wind erosion. As a result, loess deposits rapidly thin with distance from major rivers.

The sediment consists of angular particles of equal grain-size composed of quartz, feldspar, hornblende, mica, and bits of clay. The sediment is usually a buff to yellowish brown loamy deposit that is commonly unstratified due to

Figure 129 *Stratified loess, near Helena, Phillips County, Arkansas.*

(Photo by A. F. Crider, courtesy of USGS)

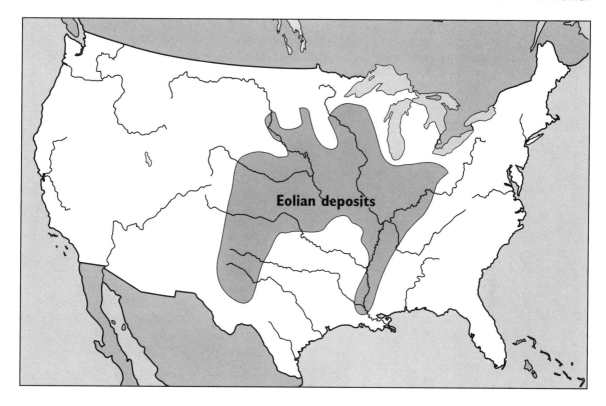

Figure 130 *Windblown soil deposits in the United States.*

a rather uniform grain size, generally in the silt size range. Loess often contains the remains of grass roots. As with mud bricks, deposits can stand in nearly vertical walls despite their weak cohesion. Loess can also cause problems in construction unless properly compacted because when wet, it tends to settle.

Loess sediments commonly occur in North America, Europe, and Asia. China contains the world's largest deposits, which originated from the Gobi Desert and attain hundreds of feet in thickness. Most loess deposits in the central United States are located adjacent to the Mississippi River Valley (Fig. 130), where nearly a quarter-million square miles are covered by sediment from the glaciated northlands. Deposits also cover portions of the Pacific Northwest and Idaho. Loess makes a yellowish fertile soil responsible for much of the abundant agricultural production of the American Midwest.

The next chapter will switch from hot deserts to cold tundra and explore some Arctic secrets.

8

ARCTIC GEOLOGY
LANDFORMS CREATED BY ARCTIC CONDITIONS

This chapter examines the various geologic forms in the Arctic and Antarctic. The Arctic regions of the world are some of the most desolate environments. Because of the molding by ice, they produce many unusual landscapes. Tundra creates a spongy, waterlogged topography widely developed in the Arctic of North America and Siberia. Unique geologic structures adorn the tundra permafrost, including boulders fashioned into polygonal shapes that have confounded geologists for centuries.

As the last remaining frontier, Antarctica is a strange land of ice. Continental glaciers bury most of its land features, with the exception of a few dry valleys, which are among the most barren deserts in the world. A significant portion of the Earth's surface is covered with glaciers that slither out of the mountains and dive into the sea, spawning armadas of icebergs.

ARCTIC TUNDRA

The Arctic is a broad region surrounding the North Pole above 66.5 degrees north latitude. It includes the Arctic Ocean and adjacent lands. It embraces all the extreme northern lands in the Arctic, including the upper portions of Alaska, Canada, most of Greenland, the northern tip of Iceland, and the northlands of Scandinavia and Russia. The Arctic Tundra of North America and Eurasia covers about 14 percent of the Earth's land surface in an irregular band winding around the top of the world north of the boreal forests and south of the permanent ice sheets (Fig. 131). Because of the tilt of the Earth's axis, the Arctic is bathed in sunlight 24 hours a day during the summer. In the winter, it is cloaked in 24-hour-a-day darkness, similar to Antarctica.

Alpine Tundra covers much of the world's mountainous terrain above the tree line and below mountaintop glaciers. While little snow precipitates in much of the Arctic, alpine areas receive abundant snowfall that precipitates at high elevations. Because Arctic tundra lies at higher latitudes, it is deprived of sunlight during the long winter months, whereas alpine tundra receives sun-

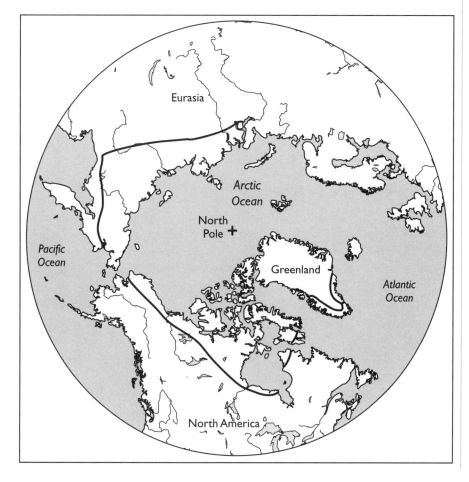

Figure 131 *The Arctic tundra line. North of this, the ground remains frozen year-round.*

light daily. The vegetation in both regions is very similar, however, consisting mostly of stunted plants often widely separated by bare rock and soil (Fig. 132).

Limited food resources, high winds, and cold temperatures during most of the year make the Arctic Tundra one of the most barren environments on Earth. Parts of the Arctic Tundra are the world's most nutrient-poor habitats. Yet algae and aquatic insects proliferate in the cold streams flowing into the Arctic Ocean. The Mackenzie River, which drains much of the Arctic region of northern Canada (Fig. 133), is one of the world's largest streams.

Plants and animals must take full advantage of a limited growing season, rainfall, and nutrient supply. The growing season generally lasts only two to three months. However, an increase in temperature from a global climate change could significantly lengthen it, dramatically altering the tundra ecology. Plants have acquired special adaptations to survive in this demanding environment, where sturdy as well as frail species live.

The hardiness of tundra species is exemplified by the seeds of the Arctic lupine collected from an ancient lemming burrow, dating near the end of the last ice age about 10,000 years ago. Perhaps the most remarkable aspect was how well these seeds were preserved. When planted, they actually germinated, blossoming with delicate flowers to become, by far, the oldest plants ever to grow.

Figure 132 Polygonal ground on a mountaintop east of Rex Dome at the northern foothills of the Alaska Range, showing plant growth along only the relatively clay-free borders, Yukon region, Alaska.

(Photo by C. Wahrhaftig, courtesy of USGS)

Figure 133 *Numerous kettles stud the tundra of Cape Bathurstand Mackenzie River area, Northwest Territory, Canada.*

(Courtesy of NASA)

Despite its pristine appearance, the Arctic Tundra is heavily polluted. The polar front—the boundary between cold polar and warm temperate air masses—is linked with the jet stream, which moves with the polar front southward in winter and northward in summer. During the winter months, when the polar front sweeps across polluted areas of the Northern Hemisphere, it removes atmospheric pollution and transports it to the tundra regions, contaminating the once pristine skies with an Arctic haze.

The haze has become a nearly permanent feature of the far north in the last half century, making the Arctic as polluted in winter and early spring as parts of the North America and Eurasia. The smog originates mostly in Europe and northwest Asia, where the Russian city of Norilsk with its vast ore-smelting complex was suspected of generating most of the pollution. The Arctic haze, which is often as bad as the air pollution in some American suburds, does its most harm by blocking out sunlight, a premium at these high latitudes. The smog persists due to a lack of significant precipitation needed to scrub the pollutants out of the atmosphere.

Acid precipitation can be quite harmful to delicate tundra flora. Lichens living in areas where no other plants exist grow very slowly because of the harsh conditions. The plants absorb dangerous substances from atmospheric moisture and dust that can build to toxic levels. Therefore, they are usually among the first to perish during environmental stress. Animals feeding on the contaminated plants can also accumulate lethal levels of the poisons.

Arctic Tundra is an extremely fragile environment. Even small disturbances cause considerable damage. Reindeer overgrazing on the sparse grassland can decimate large areas. Petroleum and mineral exploration can ruin substantial acreage, and vehicle tracks remain decades later (Fig. 134). The great northern boreal forest and Russia's taiga are a vast band of conifers and other softwoods stretching across the northlands of North America and Eurasia. Over the past century, the forests have absorbed excess carbon dioxide generated by industrial activities. Unfortunately, the absorption of this greenhouse gas has fallen off due to logging and massive increases in tree

Figure 134 *Tractor trail on the North Slope of Alaska. The small ponds are due to thawing of the permafrost in the roadway.*

(Photo by O. J. Ferrians, courtesy of USGS)

dieback from fires, acid rain, and diseases. These are generally due to warmer weather in much of the Arctic since the 1970s.

Trees that migrate northward in response to global climatic changes could cause a decline in native Arctic ecosystems. The effects of this could last for centuries. Forests would creep toward the north, and other wildlife habitats in the Arctic Tundra would disappear entirely. The northward-shifting habitats could force some tundra species into extinction from the breakup of mixed habitats into more uniform ones.

If these changes occur too rapidly, many Arctic communities would be unable to adjust fast enough to the new conditions. The flat, open tundra, where only lichens and other low-growing plants live, could witness an incursion of taller plants. These would, in turn, affect the nesting behavior of shorebirds breeding in the Arctic. Fish and marine mammals living in the Arctic waters could also be adversely affected by a climate change.

Warmer global temperatures, which are magnified in the high latitudes, might thaw the Arctic Tundra. This would release into the atmosphere vast quantities of carbon dioxide and methane, another potent greenhouse gas, which is trapped in the soil by decomposing plant material. Increased carbon dioxide also fertilizes the wetlands, making them emit more methane. These processes could produce a runaway greenhouse effect, raising global temperatures to lethal levels. As a result, the high-latitude regions of the Northern Hemisphere would witness increased cloudiness and precipitation and decreased permafrost.

Alaska's northern tundra shows evidence that a warming climate might already be releasing carbon dioxide from the land. Boreholes drilled across the Arctic Tundra of northern Alaska over a distance of 300 miles indicate anomalous warming in the upper 300 to 500 feet of permafrost and rock. The warming is between two and four degrees Celcius over the 20th century, a temperature greater than the global average warming of perhaps one degree Celsius during the same period. This increase has occurred because the effects of greenhouse warming have a greater influence in the Arctic regions.

PERMAFROST

Permafrost in the Arctic and mountainous regions underlies about a fifth the land area of the world. Most of the ground in the Arctic is frozen year-round (Fig. 135), and only the top few inches of soil thaw in the short summer season. The permafrost in Alaska varies from about 1,300 feet thick in the northern region to less than a foot thick in the south. The permafrost consists of an active layer that thaws during the summer and is underlain by hundreds of feet of frozen layers. The active layer thickness depends on the exposure, slope,

Figure 135 *A large mass of ground ice exposed by gold placer mining in silt bluff near Fairbanks, Yukon region, Alaska.*

(Photo by O. J. Ferrians, courtesy of USGS)

amount of water, and vegetation cover. These factor significantly affects the thermal conductivity of the soil.

Even though summer sunlight bathes the ground 24 hours a day, the soil temperature seldom rises much above the freezing point of water because most of the radiant energy is used to melt the soil ice. Often over a lengthy period, this freeze-thaw sequence produces bizarre patterns in the ground. The patterned ground graces the periglacial and alpine regions worldwide from Antarctica to the northern Arctic. Some features consist of regular polygonal shapes created when the ground heaves up by the expansion of ice as it freezes.

Severe subsidence called solifluction is a type of mudflow that occurs in permafrost regions of the world. It is a slow downslope movement of water-logged sediments that causes ground failures in permafrost. When frozen ground melts from the top down during spring in the temperate regions or during summer in areas of permafrost, surface mud tends to slide downslope over a frozen base. This type of ground motion can create problems with construction projects, especially in the far northern permafrost regions. Foundations must extend down to the permanently frozen layers or whole buildings could be damaged by the loss of support or by foundation slippage.

The slow downslope movement of overburden and bedrock, called creep, might occur very rapidly where frost action is prominent. After a freeze-thaw sequence, material moves as the ground expands and contracts. Frost heaving is another movement of soil material associated with cycles of freezing and thawing. Fine soil is thought to be more susceptible to frost heaving than coarse grains or rocks. Frost heaving thrusts boulders and other structures upward through the soil, often posing serious construction problems (Fig. 136).

Frost action causes mechanical weathering by exerting pressures against the sides of cracks and crevices in rocks as water freezes. The frost wedging widens the cracks, while surface weathering rounds off the edges and corners. The resulting landscape resembles numerous miniature canyons up to several feet wide carved into the bedrock.

Lying at the margins of glaciers in many areas of the northern tundra are rugged periglacial regions. A periglacial environment is defined by the conditions, processes, and topographic features of areas adjacent to the borders of a glacier. A periglacial climate results in low temperatures that fluctuate about the freezing point and strong wind action during certain times. Periglacial processes directly controlled by glaciers sculpted features along their periph-

Figure 136 *Frost-heaved piling of a bridge spanning the outlet of Clearwater Lake near Big Delta, Donnelly District, Alaska.*

(Photo by T. T. Pewe, courtesy of USGS)

ery. Cold winds blowing off the glaciers affect the climate of the glacial margins and help create periglacial conditions. The zone is dominated by such activities as frost heaving, frost splitting, and sorting. This results in immense boulder fields created out of once-solid bedrock.

POLYGONAL GROUND

In the periglacial and alpine regions of Antarctica and the Arctic Tundra, soil and rocks are fashioned into orderly patterns. As the ground thaws in the Arctic summer, the retreating snows reveal rocks arranged in a honeycomb-like network. They give the landscape the appearance of a tiled floor (Fig. 137). Even on Mars, orbiting spacecraft have detected furrowed rings, polygonal fractures, and ground ice patterns of every description, suggesting the planet's surface water has long gone underground.

The patterned ground occurs in the northern lands of North America and Eurasia where the soil is exposed to moisture and seasonal freezing and thawing cycles. The polygons range in size from a few inches across when composed of small pebbles to several tens of feet wide when large boulders form protective rings around mounds of soil. The polygons appear to have originated by similar processes that cause frost heaving, which thrusts boulders upward through the soil. This is a major annoyance to northern farmers, who find a new harvest of rocks every spring. Boulders are also known to penetrate highway pavement, and fence posts have heaved completely out of the ground.

The boulders appear to move through the soil by a pull from above and by a push from below. If the top of the rock freezes first, the expanding frozen soil pulls it upward. When the soil thaws, sediment gathers below the rock, which settles at a slightly higher level. The expanding frozen soil beneath the rock also heaves it upward. After several frost-thaw cycles, the boulder finally rests on the surface.

The patterned ground appears to form by soil of mixed composition moving upward toward the center of the mound and downward under the boulders in a convective or circular motion. This is similar to bubbles rising in a pot of boiling water. The material moves down under the gravel similar to the action of a subduction zone. The coarser material, consisting of gravel and boulders, is gradually shoved radially outward from the central area, leaving behind the finer materials. The arrangement of rocks thereby suggests the soil was churned up by convection.

The Arctic soil provides an assortment of other geometric designs, including steps, stripes, and nets between the circles and polygons (Fig. 138). They reach 150 feet in diameter. Relics of ancient surface patterns measuring up to 500 feet have been found in former permafrost regions. These, among many other mysterious features, make the Arctic one of the most fascinating places on Earth.

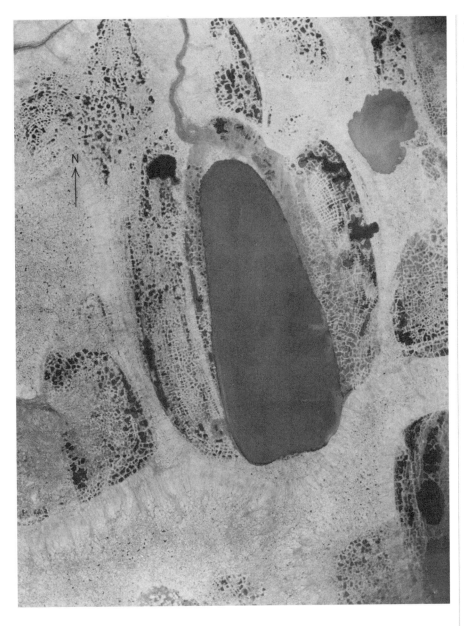

Figure 137 *Ice wedge polygons, Arctic slope, Barrow District, northern Alaska.*

(Photo by D. M. Hopkins, courtesy of USGS)

GREENLAND AND ICELAND

Continental glaciers are the largest ice sheets. They covered as much as one-third of the land surface during glacial periods. Only Antarctica and Greenland have substantial ice masses and now hold about 30 percent of the total ice volume of the last ice age. A continental glacier moves in all directions outward

Figure 138 *Stone nets with well-developed sorting due to exclusion of cobbles into a polygonal net, St. Lawrence Island, Bering Sea.*

(Photo by H. B. Allen, courtesy of USGS)

from its point of origin. It completely engulfs the land except for isolated high mountain peaks projecting above the surface of the ice. The term *ice cap* also describes a small glacier that spreads out radially from a central point in a manner similar to that on Iceland.

Greenland holds the largest ice sheet in the Northern Hemisphere. The world's biggest island separated from Eurasia and North America about 60 million years ago. About 4 million years ago, Greenland acquired a permanent ice cap two miles thick or more in some places. Evidence of dropstones on the ocean floor, which are boulders released from melting icebergs that calved off the Greenland Ice Sheet, suggests, however, that ice covered major sections of the island perhaps as early as 8 million years ago.

The accumulation of snow precipitated from storm systems traversing across the surface of the glacier nourishes the Greenland ice sheet. The loss of ice at the boundary regions adjacent to the ocean balances the glacier's growth. Large icebergs calving off glaciers entering the sea become a shipping hazard in the North Atlantic (Fig. 139 and Fig. 140).

Greenland also hosts some of the world's oldest rocks. The Isua Formation in a remote mountainous area in the southwestern region comprises metamorphosed marine sediments formed about 3.8 billion years ago. Surprisingly, Greenland is devoid of significant earthquake activity. This is supposedly because the weight of its massive ice sheet pressing down on the crust stabilizes existing faults, thus inhibiting fault slip.

Around 1,000 years ago, during the Medieval Little Optimum, when average global temperatures were one to two degrees Celsius warmer than today, the Greenland Ice Sheet shrunk sufficiently to allow the Norse to establish permanent settlements. Perhaps the frozen island was named Greenland to entice people to settle there. The climate during early colonization was unusually warm, so perhaps parts of the island were green after all.

As its name suggests, Iceland is one of the coldest inhabited places on Earth and holds the second largest ice cap in the Northern Hemisphere. The Norse who originally discovered the island witnessed energetic displays of hot water and steam gushing from the ground. The original settlers named the capital city Reykjavik, which means "smoking bay," for its wisps of steam rising from the water's surface.

Figure 139 *A Coast Guard icebreaker approaches a large tabular iceberg in Melville Bay, Greenland.*

(Photo by PH1 Briscuiti, courtesy of U.S. Coast Guard)

Figure 140 *The Greenland Ice Sheet. Dashed line shows principal iceberg calving areas.*

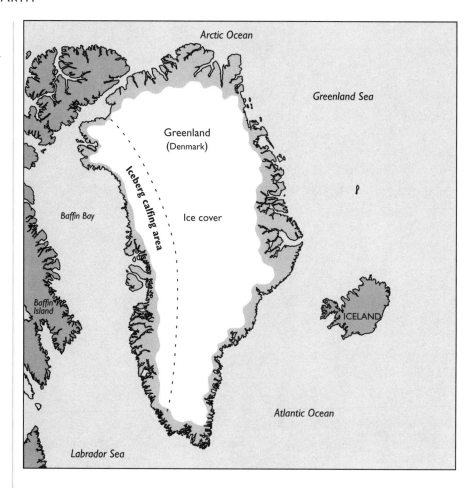

Iceland is a broad volcanic plateau of the Mid-Atlantic Ridge, which rose above the sea about 16 million years ago. The abnormally elevated topography extends some 900 miles along the ridge, 350 miles of which lies above sea level. South of Iceland, the broad plateau tapers off and forms a typical midocean ridge. A mantle plume rising from the bottom of the mantle and lying beneath the plateau apparently augments the normal volcanic flow of the Mid-Atlantic Ridge, making Iceland's existence possible.

The island is unique because it straddles a spreading ridge system, where the two plates of the Atlantic Basin and adjacent continents pull apart. A steep-sided, V-shaped valley runs northward across the entire length of the island and is one of the few expressions of a volcanic rift system on land. Many volcanoes flank the rift, which makes Iceland among the most volcanically active places on Earth.

Volcanism on Iceland produces glacier-covered volcanic peaks up to a mile high and generates intense geothermal activity. Although Iceland is fortunate to possess such an abundant supply of energy for electric power and heating, it is not without its dangerous side effects. Frequent volcanic eruptions plague the island. The most destructive eruption in modern times buried much of the fishing village of Vestmannaeyjar on the island of Heimaey in 1972 (Fig. 141). An underglacier eruption in the sparsely populated southeastern part of the country in 1996 created massive flooding as gushing meltwaters and icebergs raced 20 miles to the coast.

THE ANTARCTIC

Antarctica has geographic features similar to those of other continents. However, its mountain ranges, high plateaus, lowland plains, and canyons are buried under a sheet of ice that is, in places, up to three miles thick (Fig. 142). Antarctica is divided by a wall of mountains called the Transantarctic Range that forms the spine of the continent. It separates the eastern and western ice sheets into a large East Antarctic ice mass and a smaller West Antarctic lobe about the size of Greenland. The ice cap averages about 1.3 miles thick with a mean elevation of about 7,500 feet above sea level. Barren mountain peaks soar 17,000

Figure 141 *A lava flow that partially engulfed a building from the January 23, 1973 eruption on Heimaey, Iceland.*

(Courtesy of USGS)

Figure 142 Admiralty Mountain Range near Hallette Station, Antarctica.

(Photo by G. L. Arnald, courtesy of U.S. Navy)

feet above the ice sheet, and hurricane force winds shriek off the ice-laden mountains and high ice plateaus.

Antarctica contains about 90 percent of the Earth's ice, amounting to 70 percent of all the world's freshwater. The bulk of the ice has not changed significantly over the past 15 million years. Trapped under the thick polar ice cap is a huge expanse of water the size of Lake Ontario, covering more than 5,000 square miles to a depth of at least 1,600 feet. Water collected in bedrock pockets was prevented from freezing by geothermal heat from below and pressure from the ice above.

Sea ice around the frozen continent expands to 7.7 million square miles in winter, more than twice the size of the United States. It grows at an average rate of over 20 square miles per minute and usually attains a thickness of no more than three feet. Antarctic sea ice differs from that in tha Arctic, where most of the ocean is surrounded by land, which dampens the seas and allows the ice to grow over twice as thick. Some of the Arctic ice survives the summer, so in four years it doubles in thickness. However, in the Antarctic, powerful storms at sea churn the water and break up the ice, preventing it from growing any thicker.

Figure 143 *Dry valleys and mountains in Antarctica.*

(Photo by F. R. Bair, courtesy of U.S. Navy)

Yet despite all this ice, Antarctica is literally a desert. It has an average annual snowfall of less than two feet. This translates into roughly three inches of rain, making the continent one of the most impoverished deserts. Dry valleys (Fig. 143) gouged out by local ice sheets running between McMurdo Sound and the Transantarctic Mountains are the largest ice-free areas on the continent. They receive less than four inches of snowfall yearly, most of which blows away by strong winds. The landscapes of the dry valleys are very old. Some surfaces are quite steep and appear to have remained virtually unchanged for 15 million years.

Thick, floating ice shelves nearly the size of Texas covering the Ross and Weddell Seas dominate West Antarctica. The elevation in this region is generally low. Most of the ice rests on glacial till lying mostly below sea level. The till is a mixture of ground-up rock and water that acts as a lubricant to aid the ice sheet's slide into the ocean.

The Filchner-Ronne Ice Shelf south of the Weddell Sea, the most massive floating block of ice on Earth, has two distinct layers. The top layer measures about 500 feet thick and is composed of ice mostly formed by falling snow. The bottom layer measures about 200 feet thick and consists of frozen

seawater. The freshwater layer contains opaque and granular ice similar to the upper portion of a glacier. In contrast, the transparent marine shelf ice displays many inclusions of marine origin such as plankton and clay particles. Free-floating ice platelets recrystallize at the base of the marine layer, forming a slush that compacts into solid ice.

The East Antarctic Ice Sheet rests on solid bedrock and therefore is reasonably stable. In contrast, the ice sheet in West Antarctica rests below the sea on bedrock and glacial till. It is surrounded with floating ice pinned in by small islands buried below the ice. The ice on the mainland is unimaginably heavy and depresses the continental bedrock nearly 2,000 feet. The ice cap is so thick that the region is practically devoid of earthquakes because the great weight of the ice prevents slippage along faults.

ICE STREAMS

The West Antarctic Ice Sheet is inherently unstable. A sustained warmer climate could cause it to collapse and crash into the sea in a similar manner that North American glaciers surge into the ocean (Fig. 144). Increasing water pressure under the glacier can lift it off its bed, overcoming the friction between ice and rock. This frees the glacier, which quickly slides downslope under the pull of gravity. Rapid glacial flow develops on soft beds as water-saturated sediment weakens and no longer withstands the shear stress applied by the overlying ice. The sediment therefore begins to deform and behaves as a plastic.

Ice streams several miles broad in West Antarctica proceed down mountain valleys to the sea. The ice streams flow through stationary or slower-moving glaciers bounded on both sides by rock. The mass of ice flowing to the coast is significantly greater than that accumulating at the ice stream's source, indicating a possible instability. Some ice streams travel quite rapidly. For example, the Rutford Ice Stream cruises along at speeds of over 1,200 feet a year, roughly 100 times faster than the ice bordering it.

Rivers of solid ice slowly flow outward and down to the sea on all sides of the Transantarctic Range and then onto giant floating ice shelves. The ice streams act as pipelines carrying ice from the stable continental interior toward the ocean, where they form ice shelves up to 5,000 feet thick that break up into icebergs (Fig. 145). In early 1995, a mammoth iceberg broken off the Larsen Ice Shelf measured some 50 miles long, 25 miles wide, and 700 feet thick, almost as large as the state of Rhode Island.

Figure 145 *A giant iceberg at Hallett Station, Antarctica.*

(Photo by Cdr. Goodwin, courtesy of U.S. Navy)

The ice streams escape through mountain valleys to the ice-submerged archipelago of West Antarctica and onto the ice shelves of the Ross and Weddell Seas (Fig. 146). Normally, ice shatters under stress. However, because of its immense size, a glacier acts as a flowing viscous solid slowly creeping over the landscape. The ice flows as long as its surface runs downhill, even if the ground underneath slopes uphill. Sometimes, the ground topography below the ice funnels it into faster moving streams. The ice along the edges of the streams is softer, allowing the ice streams to slip easily through the ice sheet.

Pools of liquid water beneath the glaciers lubricate the ice streams and help them flow down the mountain valleys to the sea. Ice streams up to several miles broad glide smoothly along the valley floors at approximately half a mile per year. When a glacier surges, the water stops flowing and spreads out beneath the glacier. This watery undercoat lubricates the glacier, which allows large parts of the ice sheet to race along the ice streams at speeds several times

Figure 146 *Ice shelves in Antarctica are shown in the blackened areas.*

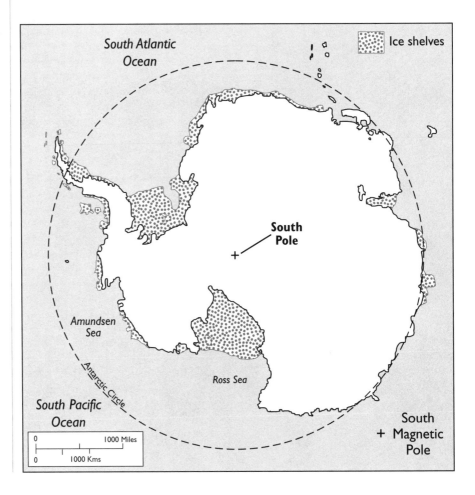

faster than usual. The glaciers gliding along on the valley floor eventually plunge into the sea (Fig. 147).

The ice streams are marked by deep crevasses, which are cracks or fissures resulting from stress due to movement in the ice. They are generally several tens of feet wide, 100 or more feet deep, and up to 1,000 or more feet long. Deep crevasses often flank the banks of glaciers, where they contact the walls of the glacial valley. Crevasses also run parallel to each other down the entire length of the ice streams, especially when the central portion of the glacier flows faster than the outer edges. Sometimes a stream of meltwater can be heard gurgling far below from open crevasses slicing through the glacier. Snow bridges occasionally span the crevasses. In some cases, these bridges completely hide the crevasses from view, making them extremely treacherous.

After discussing ice, the next chapter will examine what it can do to the landscape.

Figure 147 *Strap penguins on ice floes in Arthur Harbor, Antarctica.*

(Photo by G.V. Graves, courtesy of U.S. Navy)

9

GLACIAL LANDSCAPES
LANDFORMS CREATED BY GLACIERS

This chapter examines the different kinds of landforms resulting from glacial erosion during the ice ages. Many northern lands owe their unique landscapes to immense glaciers that swept down from the polar regions and trampled everything in their paths during the ice ages (Table 10). Thick sheets of ice enveloped the northern continents, and alpine glaciers grew on nearly every mountain peak. Their legacy remains as deeply eroded rock in the high ranges of the world. The ice ages left an unusual collection of glacial structures, including cirques, kettle holes, glacial lakes, flood-ruptured ground, and many other landforms sculpted by roving ice sheets.

Thick glacial deposits called erratics, tillites, and moraines covered many areas. Glacial sediments buried older rocks, forming elongated hillocks called drumlins. Glacial debris from outwash streams formed long, sinuous sand deposits called eskers. Mounds of glacial sand piled up into kames. Finer sediment falling to the bottoms of glacially fed lakes formed varves, whose distinct layering aids in correlating the various glacial episodes.

TABLE 10 THE MAJOR ICE AGES

Number of years ago	Event
10,000–present	Present interglacial
15,000–10,000	Melting of ice sheets
20,000–18,000	Last glacial maximum
100,000	Most recent glacial episode
1 million	First major interglacial
3 million	First glacial episode in Northern Hemisphere
4 million	Ice covers Greenland and the Arctic Ocean
15 million	Second major glacial episode in Antarctica
30 million	First major glacial episode in Antarctica
65 million	Climate deteriorates, poles become much colder
250–65 million	Interval of warm and relatively uniform climate
250 million	The great Permian ice age
700 million	The great Precambrian ice age
2.4 billion	First major ice age

GLACIAL EROSION

Glaciers are among the most effective agents of erosion, especially in mountainous regions of the world with vast expanses of exposed rock upon which to work (Fig. 148). During the ice ages, continent-wide glaciers buried many mountains in the northlands of Eurasia and North America, where glacial ice linked the Rocky Mountains with ranges in northern Mexico. In the Southern Hemisphere, small ice sheets expanded in the mountain ranges of Australia, New Zealand, and the Andes of South America. Throughout the world, alpine glaciers capped mountains that are presently ice-free. Massive glaciers excavated some of the most monumental landforms. Outwash streams from glacial meltwater carved out many peculiar landscapes.

The glaciers descended from the mountains and spread across most of the northern lands, running over the terrain like an icy bulldozer. The power of glacial erosion is well demonstrated by the presence of deep-sided valleys carved out of mountain slopes (Fig. 149) by flowing ice a mile or more thick. Glacial erosion radically modified the shapes of stream valleys occupied by glaciers. The process is most active near the head of a glacier, where the ice deepens and flattens the gradient of the valley.

Erosion with glacial ice acting as the primary agent reduces the land surface by glacier-transported rock fragments, bedrock scouring, and the erosive action of meltwater streams. Most glacial erosion involves the removal of rock by plucking. Abrasion smooths and polishes the resulting form. Small hills or knobs in valleys overridden by glaciers become rounded and smoothed by abrasion.

A glacier moving over bedrock abrades it by glacial scouring from the action of grinding or rasping. The abrasive agent is rock material dragged along by the glacial ice. Because of its fluid nature, ice does not erode the rock itself. Instead, it plucks fragments of bedrock by the plastic flow of the ice. The rocks then become part of the moving glacier. Boulders embedded in the ice gouge deep cuts into the easily eroded bedrock. Smaller rocks cut parallel striations or scratches in the bedrock (Fig. 150), and finer material polishes it to a smooth finish.

A remarkable example of the power of glacial erosion is a 200-mile-long by 70-mile-wide delta-shaped landform in eastern South Dakota known as the Coteau des Prairies. It is a low-relief formation created from a deposit of hard quartzite (metamorphosed quartz sandstone) that split the southward-flowing glacier of the last ice age into two lobes. The ice scoured the lowlands on either side but did not overtop the Coteau itself, leaving it standing alone above the adjacent terrain.

At the end of the last ice age, massive floods raged across the land as water gushed from huge reservoirs trapped beneath the melting glaciers. Water

Figure 148 Cutbank Pass, Glacier National Park, Montana.

(Photo by M. R. Campbell, courtesy of USGS)

flowing under the ice surged in vast, turbulent sheets that scoured deep grooves in the crust, forming steep ridges out of solid bedrock. Torrents of meltwater laden with sediment surged along the Mississippi River to the Gulf of Mexico, significantly widening its channel and depositing new soil. Many other rivers overreached their banks to carve out broad floodplains.

Figure 149 *Glaciated valley at Glacier National Park, Montana.*

(Photo by G. A. Grant, courtesy of National Park Service)

GLACIATED VALLEYS

A glacial valley is a river valley that was glaciated during the Ice Age. Glaciers did not cut the original valley but only altered it by converting the formerly V–shaped valley into one that is U shaped (Fig. 151). This U shaped valley has a broad, flat bottom as much as 1,000 feet or more deep. A glacier tends to straighten the valley it erodes because ice cannot turn as sharply as a river due to its higher viscosity.

Figure 150 *Glacial polish, striae, and grooves in the Upper Kern Canyon, Sequoia National Park, Tulare County, California.*

(Photo by F. E. Matthes, courtesy of USGS)

Figure 151 *A U-shaped glaciated valley, Red Mountain Pass, south of Ouray, Colorado.*

(Photo by L.C. Huff, courtesy of USGS)

Glacial erosion removes ridges on the insides of curves within the stream valley and eliminates projecting spurs, which are ridges extending laterally from a mountain range. The glaciated valley floor is often irregular because ice more readily erodes fractured or weak rock. As a result, giant steps form at intervals along the length of the valley.

The principal river valley, whose source is near the mountain crest, contains a greater volume of ice than a tributary stream valley. Consequently, the glacial ice erodes the main valley deeper than the tributary valley. After the ice melts, the tributary stream flows through a hanging valley above the main stream, into which it pours from a waterfall (Fig. 152).

Glaciers at least a mile thick buried many valleys during the Ice Age. As the glacial ice extended far down the valleys, it ground rocks on the valley floors. Rivers of solid ice with embedded rocks moved along the valley, grinding down the bedrock like a giant file as glaciers advanced and receded. The overriding ice also inscribed parallel furrows or striations on the valley floors as they sliced down the mountainsides.

Figure 152 *Nevada Fall at the head of Yosemite Valley, Yosemite National Park, California.*

(Photo by G. K. Gilbert, courtesy of USGS)

Glacial striae cover large areas of northern Eurasia and North America. They are finely cut, nearly parallel grooves or scratches cut into the bedrock surface by rock fragments carved out by glaciers. The striae are also engraved on the transported rocks and are excellent indicators of the direction of glacial flow. Glacial striae are usually found on Pleistocene deposits. However, they also appear on rocks from earlier glaciations as far back as the Precambrian era.

Fjords are long, narrow, steep-sided inlets in glaciated mountainous coasts. During the last ice age, glaciers gouged deep fjords out of the coastal mountains of Norway, Greenland, Alaska, British Columbia, Patagonia in southern South America, and Antarctica. As a tidewater glacier on the coast erodes its valley floor below sea level, it cuts a steep-walled, troughlike arm of the ocean. When sea levels returned to normal at the end of the Ice Age, the ocean invaded deeply excavated glacial troughs in the coastline. The sidewalls along the fjord are characterized by hanging valleys and tall waterfalls.

CIRQUES AND ARÊTES

Alpine glaciers flowing down mountain peaks gouged out large pits called cirques (Fig. 153), which is from the French, meaning "circle." They are semicircular basins or indentations with steep walls high on a mountain slope at the head of a valley. Cirque walls are cut back by the disintegration of rocks lower down the mountainside. The rock material embedded in the glacier gouges a concave floor, which might contain a small mountain lake called a tarn.

The expansion of adjacent cirques by glacial erosion creates arêtes, horns, and cols. An arête, from the French word meaning "fish bone," is a sharp-crested, serrated, or knife-edged ridge that separates the heads of abutting cirques. It also forms a dividing ridge between two parallel valley glaciers. In Glacier National Park, Montana, the Continental Divide—the boundary separating river systems flowing in opposite directions—follows an impressive arête called the Garden Wall. A col is a sharp-edged or saddle-shaped pass in a mountain range formed by the headword erosion where cirques meet or intercept each other. When three or more cirques erode toward a common point, they form a triangular peak called a horn (Fig. 154). The Matterhorn in the Swiss Alps is perhaps the world's best example of this type of formation.

GLACIAL LAKES AND KETTLES

Glacial lakes excavated by roving glaciers dot many northern lands. Lakes over much of Canada and the northern United States formed when glaciers eroded large depressions in the bedrock. Lake Agassiz in southern Manitoba was

Figure 153 *A hanging cirque west of Siyeh Glacier, Glacier National Park, Glacier County, Montana.*

(Photo by H. E. Malde, courtesy of USGS)

once a huge lake, larger than any of the existing Great Lakes. Its waters were trapped in a topographic depression and held back by a rise called a rock sill.

Similar large reservoirs of meltwater included Lake Lahontan in western Nevada, which expanded 10 times wider than today's remnant. In addition, Lake Bonneville, which covered over 20,000 square miles of Utah and Nevada, is now a dry salt flat punctuated by the Great Salt Lake, the only remaining body of water. The salt pan of Death Valley is the evaporated remains of a series of lakes. The largest was Lake Manly, whose basin filled with runoff from the glaciated Sierra Nevada Mountains between 75,000 and 10,000 years ago.

The Great Lakes on the border between the United States and Canada are the largest glacial lakes in the world. Often, high lake levels result in severe shore erosion. Sudden barometric pressure changes on the lakes cause water to slosh back and forth. This produces a large wave called a seiche, which can seriously erode the shoreline. Material eroded from the continent flows into the lakes, and the continuous buildup of sediment gradually makes them shallower. As the lakes continue to fill with sediment, they will eventually become dry, flat, featureless plains until the glaciers return to scour out their basins.

The glaciated areas of North America contain tens of thousands of pits called kettle holes (Fig. 155) that originated from large blocks of ice buried by glacial outwash sediments. The depressions are circular or elliptical because ice blocks tend toward roundness as they melt. A glaciere is a deeply buried block of ice that remains frozen year-round even in temperate climates. The underground ice formation sometimes forms a cavity, where an ice mass is kept from melting throughout the year.

Most kettles occur singly or in groups. They measure up to several miles wide and 100 feet or more deep. A collection of large numbers of kettles pro-

Figure 154 *The Matterhorn-like crest of Three Fingered Jack resulted from glacial erosion of a large volcanic cone, Mount Jefferson Primitive area, Jefferson County, Oregon.*

(Photo by W. G. Walker, courtesy of USGS)

Figure 155 *A kettle hole near Buckfield, Oxford County, Maine.*

(Photo by F. G. Clapp, courtesy of USGS)

duces a terrain in the shape of basins and mounds, called knob–and–kettle topography. The landscape features an irregular assemblage of knolls, mounds, or ridges between depressions or kettles and sometimes contains swamps or ponds. The undulating landform is a type of terminal moraine, possibly created by slight oscillations of an ice front as it recedes. A section of knob–and–kettle topography called hummocky moraine develops either along a live ice front or around masses of stagnant ice.

GLACIAL ERRATICS

Many mountainous areas once covered by glaciers display massive blocks of granite weighing several thousand tons strewn across the land (Fig. 156). The presence of these erratic glacial boulders helped lead to the recognition of widespread continental glaciation. In northern Europe, huge boulders laid scattered about as though simply dumped there. Apparently, as the glaciers melted, large boulders were lowered down on top of smaller rocks embedded in the ice below. Similar boulders spread across the slopes of the Jura Mountains in Switzerland can be traced back to the Swiss Alps over 50 miles away. Similarly, glacially deposited boulders in other parts of the world can be followed back to their places of origin.

Figure 156 *Erratic glacial boulder in the Sierra Nevada Range, Fresno County, California.*

(Photo by G. K. Gilbert, courtesy of USGS)

The boulders were originally called drift because they appeared to have drifted in by flowing water or on floating ice. Today, the term *glacial drift* refers to all rock material deposited by glaciers or glacier-fed streams and lakes. The greatest thicknesses of rock occur in buried valleys (Fig. 157). Drift is divided into two types of material. One is till deposited directly by glacial ice and shows little or no sorting or stratification (layering) as though dumped haphazardly. The other is stratified drift, which is a well-sorted, layered material transported and deposited by glacial meltwater. Streams flowing from melting glaciers rework part of the glacial material, some of which is carried into standing bodies of water. These form banded deposits called glacial varves.

Erratics are glacially transported boulders embedded in glacial till or exposed on the surface. They range in size from pebbles to massive boulders. Erratics have traveled as far as 500 miles or more. Erratics composed of distinctive rock types can be traced to their places of origin. They indicate the direction of glacial flow. Indicator boulders are erratics of known origin used to locate the source area and the distance traveled for any given glacial till. Their identifying features include a distinctive appearance, unique mineral assemblage, or characteristic fossil continent. For example, erratics scattered from Iowa to Ohio contain native copper torn from an outcrop in northern Michigan.

The erratics are often arranged in a boulder train. They form a line or series of rocks originating from the same bedrock source and extend in the direction of glacial movement. A boulder fan is a conically shaped deposit containing distinctive erratics derived from an outcrop at the apex of the fan. The angle at which the margins diverge measures the maximum change in the direction of glacial motion.

Icebergs appear to have deposited certain erratics in nonglacially transported marine sediments. Large, out-of-place boulders strewn across the central desert of Australia far from their sources suggests they rafted out to sea on slabs of drift ice when a large inland waterway invaded the continent during the Cretaceous period. When the icebergs melted, the huge rocks dropped to the ocean floor, where their impacts disturbed the underlying sediment layers. When Australia drifted into the subtropics, the central portion of the continent became a large desert. Lying in the middle of the desert sands are curious-looking boulders of exotic rock called dropstones that measured as much as 10 feet across and originated from great distances away.

Figure 157 *Exposure of Pleistocene glacial deposits in a stream bank along the Sturgeon River, Baraga County, Michigan.*

(Photo by W. F. Cannon, courtesy of USGS)

Evidence of ice-rafted boulders also exists in glacial soils in other parts of the world, including the Canadian Arctic and Siberia. Their presence suggests that the climates of the high latitudes were cold enough for ice to form even during one of the warmest periods in geologic history. Similar boulders were found in sediments from other warm periods as well. Even today, the same ice-rafting process continues in the Hudson Bay area.

TILLITES AND MORAINES

When thick ice sheets overran much of the upper midwestern and northeastern United States during the last ice age, glacially derived sediments covered large parts of the landscape. They buried older rocks under thick layers of till, also known as boulder clay. Till is nonstratified or mixed material comprising clay and boulders directly deposited by glacial ice. The boulders are generally angular with sharp edges because they experienced little or no river transportation, which abrades and rounds rocks.

Basal till was usually laid down under a glacier. Ablation till on or near the surface of a glacier was deposited when the ice melted. The surface cover of sediment possibly protected the glacier from the heat of the sun. Some sun-heated rocks on the surface might have sunk into the glacier, forming deep depressions in the ice.

Tillites are sedimentary rocks formed by the compaction and cementation of glacial tills. They are a mixture of boulders, pebbles, and clay deposited by glacial ice and made into solid rock. Thick sequences of Precambrian tillites exist on every continent (Fig. 158). They are found among Precambrian rocks in North America, Norway, Greenland, China, India, southwest Africa, and Australia. The various sediment layers provide evidence for a series of ice ages during the Precambrian.

Permian marine sediments in Australia are interbedded with alternating layers of glacial deposits. Tillites separated by seams of coal indicate that periods of glaciation were interrupted by warm interglacial spells when extensive forests grew. The Karroo Series in South Africa is a thick sequence of late Paleozoic tillites interbedded with coal beds extending over an area of several thousand miles. Between layers of coal exist fossil plant leaves of the extinct fern glossopteris, whose restriction to the southern continents provides some of the best evidence for the supercontinent Gondwana.

The simplest of all glacial landforms are moraines (Fig. 159). They are accumulations of rock material carried by a glacier and deposited in a regular, usually linear pattern that makes a recognizable landform. The sediment ranges in size from sand to boulders and shows no sorting or bedding nor-

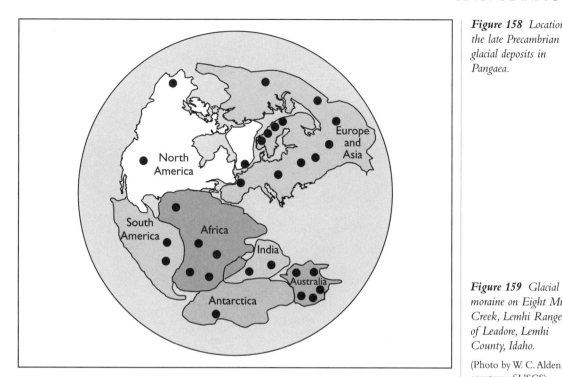

Figure 158 *Location of the late Precambrian glacial deposits in Pangaea.*

Figure 159 *Glacial moraine on Eight Mile Creek, Lemhi Range west of Leadore, Lemhi County, Idaho.*

(Photo by W. C. Alden, courtesy of USGS)

mally associated with flowing water. The rocks are generally faceted (planed off) and striated or scratched by abrasion during glacial transport.

Moraines are named according to their position in relation to the glacier. A ground moraine is an irregular carpet of till mainly composed of clay, silt, and sand deposited under a glacier. It is the most prevalent type of continental glacial deposit. A glacier overloaded with sediment at the base might drop some of its rock along the floor and ride over it. As it retreats, the glacier leaves additional rock waste, usually encased in sandy clay. Blocks of ice buried in the ground moraine leave depressions that become small ponds when the ice melts.

A terminal moraine (Fig. 160) is a ridge of erosional debris deposited by the melting forward margin of a glacier that paused long enough for till to accumulate. It is a ridgelike mass of glacial debris formed by the foremost glacial snout and deposited at the outermost edge of the glacial advance. Because of the curvature of the glacier's snout, the terminal moraine curves down valley and might form lateral moraines up the sides. A string of terminal moraines creates a broken line of irregular hills stretching along the former edge of the North American ice sheet from Cape Cod to the Rocky Mountains.

A fluctuation in the position of the glacial ice produces parallel lobes of moraine. A recessional moraine is a secondary terminal moraine deposited during a temporary halt in the retreat of a glacier. Thus, a series of recessional moraines shows the history of glacial retreat. The moraines appear as

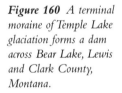

Figure 160 *A terminal moraine of Temple Lake glaciation forms a dam across Bear Lake, Lewis and Clark County, Montana.*

(Photo by M. R. Mudge, courtesy of USGS)

Figure 161 *Yentna Glacier, Mount McKinley National Park, Alaska, showing a medial moraine.*

(Photo by N. Herkenham, courtesy of National Park Service)

though bulldozed into place. However, glaciers are poor at pushing objects forward and, instead, prefer running over them. They act more like conveyor belts picking up debris beneath them, transporting the embedded sediment hundreds of miles, and dropping it out at the front edge of the glacier as it melts.

Lateral moraines are debris derived from erosion and from avalanches cascading down the valley wall onto the edge of a glacier. Avalanches are snowslides that usually begin with a mass of firmly compacted snow resting on a steep bank of weaker snowpack and often triggered by vibrations that loosen unstable layers. Lateral moraines form long ridges when the glacier recedes. The sides of a valley glacier are outlined by ridges of rock debris torn from the valley walls and fallen from the cliffs above.

When two valley glaciers merge into a single ice stream, the adjacent lateral moraines combine into a central dark streak called a medial moraine (Fig. 161). It is an enlarged zone of debris formed when lateral moraines join at the intersection of two glaciers. After the glaciers retreat, the medial moraine remains as a ridge running approximately parallel to the direction of ice movement.

DRUMLINS AND ESKERS

In parts of the high latitudes visited by glaciers, thick deposits of glacial till created elongated hillocks pointing in the same direction called drumlins (Fig. 162). They rarely stand alone but, instead, accumulate in fields often containing thousands of glacier-made hills. The hills occur in concentrated fields in North America, Scandinavia, Britain, and other areas once covered by glacial ice. Drumlin fields might comprise as many as 10,000 knolls aligned together, resembling rows of eggs lying on their sides.

The drumlin fields generally occur within a narrow band set back a short distance from a terminal moraine. The drumlins formed under the margins of glaciers or by the erosion of older moraines after reglaciation. Most drumlins consist of clayey till, some have bedrock cores, and others contain sand and gravel sediments.

The long axis of a drumlin aligns approximately parallel to the direction of ice movement. Drumlins are typically steepest and highest at the end facing the advancing ice sheet. They gently slope downward in the direction of movement of a huge lobe of the continental ice sheet, which splayed out as it plowed southward. As a result, drumlins are tall and narrow at the upstream end of the glacier and slope to a low, broad tail at the downstream end.

Figure 162 *Drumlins in Saskatchewan Province, Canada.*

(Photo by W. G. Pierce, courtesy of USGS)

Drumlins are the least understood of all glacial landforms. They appear to have originated when ice sheets contorted or deformed wet sediments lying beneath their bases. The sediments in the interior of a drumlin often form complex, swirling layers. This indicates that moving ice stretched and sheared them. How they attained their characteristic oval shape remains a mystery. Perhaps the extensive drumlin fields of North America were created during cataclysmic flood processes during melting of the vast ice sheets.

Roches moutonnées, from the French meaning "fleecy rock," are similar to drumlins. The term is applied to glaciated outcrops because they resemble the backs of sheep, thus prompting the name sheepback rock. They have a glaciated bedrock surface with asymmetrical mounds of varying shapes. The up-glacier side has been glacially scoured and smoothly abraded. The down-glacier side has steeper, jagged slopes. These resulted from glacial plucking, an erosional process by which glaciers dislodge and transport fragments of bedrock. The fragments might have been pried loose by the plastic flow of the ice around them and became part of the moving glacier.

The ridges dividing the two sides of a roche moutonnée are perpendicular to the general flow of the ice sheets. Such landforms are characteristic of glaciated Precambrian shields in the interiors of the continents. Many shields are fully exposed in areas that were ground down by flowing ice sheets during the Pleistocene Ice Ages.

Eskers (Fig. 163), from the Celtic meaning "mountain ridge," are readily recognizable glacial deposits. They are long, narrow, sinuous or straight ridges consisting of poorly sorted and stratified glacial meltwater deposits of sand and gravel. Eskers are generally shaped into winding, steep-walled ridges that extend noncontinuously up to 500 miles long but seldom exceed more than 1,000 feet wide and 150 feet high. Well-known esker areas exist in Maine, Canada, Sweden, and Ireland.

Streams that flowed through tunnels beneath or within an ice sheet created eskers in areas once occupied by the ground moraine of a continental glacier. When the ice melted, the old stream deposits remained standing as a ridge. Eskers appear to have been deposited in channels beneath or within slow-moving or stagnant glacial ice. Their general orientation runs at right angles to the glacial edge. At the margin of a glacial lake, they might form river deltas. Some eskers originated on the ice and contained ice cores.

The long, winding sand deposits were formed by glacial debris carried out from beneath the ice by outwash streams. The meltwater flowing out of a glacier is usually milk colored due to suspended fine material called rock flour produced by glacial ice abrasion. The material resembles clay and consists of fine mineral fragments that form glacial varves when meltwater flows into a lake below a glacier.

Figure 163 *An esker near Mallory, Oswego County, New York.*

(Photo by C. R. Tuttle, courtesy of USGS)

Figure 163 *An esker near Mallory, Oswego County, New York.*

(Photo by C. R. Tuttle, courtesy of USGS)

KAMES AND VARVES

Kames (Fig. 164), which come from the Scottish meaning "crooked and winding," are mounds chiefly composed of stratified sand and gravel formed at or near the snout of an ice sheet or deposited at the margin of a melting glacier. Like eskers, kames occur in areas where large quantities of coarse material are available during the slow melting of glacial ice. Meltwater must be present in sufficient quantities to redistribute the debris and deposit the sediments at the margins of the decaying ice mass.

Most kames are low, irregularly conical mounds of roughly layered glacial sand and gravel that often occur in clusters. They accompany the terminal moraine region of both valley and continental glaciers. They appear to represent sediment fillings of openings in stagnant ice. Many kames probably formed when streams flowed off the tops of glaciers onto the bare ground, dropping sediment into piles.

Some kames formed in moulins, which are holes bored through the ice to the bottom of the glacier. When meltwater from the top of the glacier plunged into the moulin, it deposited a load of sediment that piled up into a cone. Other kames appear to have formed where glacial streams escaped from the ice. Streams flowing between the sides of a glacier and the enclosing valley walls formed kame terraces that stand above the valley floor after the ice has melted.

Streams of meltwater flowing from beneath a glacier are heavily ladened with sediment rapidly deposited in a complex braided pattern of channels (Fig. 165). The braided streams spread out the debris in a series of alluvial fans that coalesce into a flat, broad, outwash plain sometimes pockmarked with kettles. The irregular surface of the terminal zone of a continental glacier produces a kame-and-kettle topography with alternating hills and hollows. The terrain also might take the form of hummocky ground peppered with knolls or knobs.

The swift-flowing stream of meltwater carries a heavy sediment load deposited at the mouth of a glacier. This forms glacial varves, which are alternating layers of sand and silt laid down annually in a lake below the outlet of a glacier. Varves generally develop on the floors of cold freshwater lakes fed by intermittently flowing glacial meltwater streams. They rarely occur in salty or brackish water. Each summer when the glacial ice melts, turbid meltwater discharges into the lake and sediments settle out differentially. The coarsest land on the bottom first, forming banded deposits.

Glacial varves, also called rhythmites, are regularly banded deposits developed by cyclic sedimentation. They have fine-grained, dark laminae alternating with coarse-grained lighter layers. Each individual pair of laminae, called a couplet, represents a one-year cycle of fast melting during summer

Figure 164 *A kame in Anchorage District, Cook Inlet region, Alaska.*

(Photo by R. D. Miller, courtesy of USGS)

Figure 165 *Braided stream on Yahtse River Delta, Yakataga district, Alaska.*

(Photo by J. H. Hartshorn, courtesy of USGS)

and slow melting during winter. Therefore, they can be used for dating purposes, especially for the late Pleistocene, when a succession of ice ages came and went almost like clockwork.

The final chapter will explore some of the many unique landforms the Earth has to offer.

10

UNIQUE LANDFORMS

LANDFORMS CREATED BY UNUSUAL GEOLOGIC ACTIVITY

A narrative about landforms would not be complete without discussing our planet's many unique landscapes. These were fashioned by diverse geologic processes, including uplift and erosion, catastrophic collapse, and tectonic activity. These forces are well represented by tall monuments of stone that dot the Earth's surface, great monoliths representing lone stands of rock scattered across the land, and grand archways and natural bridges carved from solid rock.

Much of the land's surface overlies cavities in the crust where the ground has collapsed. The Earth hosts a variety of volcanic structures, including columnar jointing, lava tubes, and craters in the strangest places. The melting glaciers at the end of the last ice age unleashed massive floods that carved out a tortured terrain as raging meltwaters raced to the sea. The presence of these geologic phenomena is a tribute to the powerful tectonic forces responsible for creating a staggering assortment of unusual landforms.

STONE MONUMENTS

Pillars of stone scattered throughout the American West were created when a resistant cap rock of hard sandstone protected the sediments below while the surrounding landscape eroded away, adorning the landscape with tall stone columns. Nowhere is the development of this phenomenon more pronounced than perhaps Monument Valley (Fig. 166) in the Four Corners region on the border of Arizona and Utah. Isolated or groups of pinnacles rise a thousand feet or more above the ground. They are not crowded together as in other parts of the mile-high Colorado Plateau. Instead, the spires, mesas, and ragged crags are widely distributed across the desert floor.

On a broader scale, erosion has whittled away flattop mountains. A remnant of the original peneplain, which literally means "almost a plain," is protected by a more resistant layer of sandstone. Many tablelands, such as the Grand

Figure 166 *Monument Valley from the top of Hunts Mesa, Navajo County, Arizona.*

(Photo by I. J. Witkind, courtesy of USGS)

Figure 167 *Erosional forms of Wasatch Formation at Bryce Canyon National Park, Garfield County, Utah.*

(Photo by H. E. Gregory, courtesy of USGS)

Mesa in western Colorado, owe their existence to a capping layer of hard basalt extruded by massive lava flows during a period of extensive volcanic activity.

Uplift and erosion have carved out fantastic forests of pinnacles, spires, and columns by a maze of ravines in Bryce Canyon National Park in southwest Utah (Fig. 167). They were fashioned from colorful rocks of the Wasatch Formation. Similarly colored sediments are responsible for the Painted Desert of eastern Arizona and the Badlands of southwestern South Dakota (Fig. 168). Numerous small streams eroded short, steep slopes and formed a unique drainage network.

In the Canyonlands of Utah and other parts of the western United States, isolated pinnacles stand out from the rest of the formation. Sometimes the statues are adorned with wide cap rocks called Mexican hats. Possibly the most prominent of these rocks lies in southeast Utah north of Monument Valley outside a small town appropriately named Mexican Hat. The structure is attributed to a resistant cap rock shaped like a sombrero sitting precariously atop an eroded remnant.

Figure 168 *Rugged out-crops of the Wasatch Formation, Badlands National Park, South Dakota.*

(Photo by J. J. Palmer, courtesy of National Park Service)

Many monuments are testaments to the persistent action of erosion that carved out lone stands of ancient rock. El Capitan Peak in Guadalupe Mountains National Park in West Texas is a massive block of limestone rising high above its sloping flanks. Sometimes a long, solitary pinnacle stands out from its surroundings such as Chimney Rock National Historic Site near Scotts Bluff, Nebraska. The area was named for a solitary, 800-foot-high bluff jutting upward in the middle of the prairie. The small town of Chimney Rock in southwest Colorado takes its name from a tall spire standing nearby.

MONOLITHS

The Colorado Plateau centered around the Four Corners region of Arizona, Utah, Colorado, and New Mexico is characterized by spectacular high-desert scenery, including tall mesas, deep canyons, and broad valleys. The region also contains great inselbergs, which comes from the German meaning "island mountains." These are isolated residual uplands standing separately above the general level of the surrounding plains.

Inselbergs are similar to the monadnocks that grace the eastern United States, where lone mountains, hills, and peneplains are the sole highlands. Inselbergs are erosional remnants rising abruptly above the ground. They are represented by ridges, domes, or hills. The sharp transition apparently resulted from anomalous weathering patterns related to the rock structure of a particular remnant as well as the topography of the area.

The regional structure of the Four Corners is predominantly horizontal. Sedimentary strata are locally deformed into broad domes, basins, and monoclines, which are steeply inclined sedimentary strata in an area where the bedding lies relatively flat. Waterpocket Fold near Lake Powell, in southeastern Utah (Fig. 169), is among the best examples of a monocline. The flanks of many monoclines present rows of flatirons that stand out from the rest of the formation due to differential weathering of the strata. They resemble old-

Figure 169 *Circle Cliffs upwarp with steeply dipping beds of the Waterpocket Fold, Garfield County, Utah.*

(Photo by R. G. Luedke, courtesy of USGS)

fashioned steam irons standing side by side. Flatirons are generously displayed in the steeply folded terrain of Utah, Wyoming, and Colorado.

Folded strata also formed by the upheaval of the crust due to the process of salt tectonics, during which a pipelike plug of salt moved upward through the sedimentary layers. The salt deposits originating from ancient brine pools buried in the crust are less dense than the surrounding rocks. The deposits slowly rise toward the surface, bulging the overlying strata upward (Fig. 170). Because salt domes make ideal structural traps for petroleum, oil and gas often collect in them. Therefore, geologists look long and hard for these deposits. Upheaval Dome near the confluence of the Colorado and Green Rivers in Canyon Lands National Park, Utah is perhaps the most striking example of a salt plug that heaved the overlying strata upward into a huge bubblelike fold three miles wide and 1,500 feet high.

Figure 170 *Sedimentary dome near Sinclair, Wyoming.*

(Photo by J. R. Balsely, courtesy of USGS)

An alternate interpretation holds that the structure conforms to the expected properties of a deeply eroded astrobleme, which is the remnant of an ancient impact structure gouged out by a large cosmic body such as an asteroid or comet striking the Earth. Erosion has removed as much as a mile or more of the overlying beds since the meteorite smashed into the ground between 30 and 100 million years ago, making Upheaval Dome possibly the planet's most deeply eroded impact crater. The original crater apparently made a 4.5-mile-wide hole in the ground that has been heavily modified by deep erosion over the many years. The dome itself appears to be a central rebound peak formed when the ground heaved upward by the force of the impact. The meteorite measured about 1,700 feet wide and crashed to Earth with a velocity of several thousand miles per hour. On impact, it created a huge fireball that would have incinerated everything for hundreds of miles.

COLUMNAR JOINTING

Long, parallel, polygonal columns produced by columnar jointing occur most frequently in basalts as well as in other extrusive and igneous rocks. During the cooling process, a reduction in volume causes a mass of molten rock to develop fractures that divide basaltic lava flows into prismatic columns with hexagonal cross sections similar to honeycombs. The columnar-joint patterns evolved from mostly tetragonal shapes to mostly hexagonal ones as the joints grew inward from the flow surfaces as cooling proceeded. The regularity and symmetry of the fracture patters have long fascinated geologists.

Early legends have attributed the formation of columnar joints to supernatural phenomena as reflected by the names of many sites bearing these features. Devils Tower (Fig. 171) in northeast Wyoming is an eroded volcanic plug of solidified magma that filled the main vent of an extinct volcano, rising 1,300 feet above the surrounding prairie. Erosion has left the more resistant rock standing alone in stark contrast to the typical western landscape. Along its flanks, columnar jointing formed by the shrinking of the magma as it cooled, creating fractures that run through the entire length of the plug.

Similarly, in northwest New Mexico, a jagged monument called Shiprock (Fig. 172) rises 1,400 feet above the mostly flat terrain. The volcanic neck is the remnant of volcanic eruptions that occurred over 30 million years ago. Three large dikes radiate outward like spokes of a giant wheel. Because the dike rocks are more resistant than the surrounding material, they formed long ridges when exposed by erosion.

Columnar jointing is also prominent at Devils Postpile (Fig. 173) southeast of Yosemite National Park in east-central California. The area contains rolls upon rolls of six-sided columns in a massive lava flow. As the lava cooled

and shrank, it developed cracking and jointing, which shoot vertically through the thick lava flow. Volcanic columns also stretch across the bare cliffs of the Columbia River Plateau, which covers much of Washington, Oregon, and Idaho in basalt floods several thousand feet deep.

ARCHES AND NATURAL BRIDGES

Arches and natural rock bridges are among the most fascinating geologic features. Across many arid lands stand magnificent stone archways (Fig. 174). The term *arch* refers to a span of rock that has no natural stream flowing beneath it. Arches are especially plentifully in Arches National Monument near Moab, Utah and delight a large number of visitors to the park. The arches formed when rock eroded at different rates due to a variance in resistance to erosional forces. They were created partly by wind erosion of thick sandstone beds. Rainwater first loosened the sand near the surface, while wind removed the loose sand grains. Wind erosion then abraded the rock, cutting through it in a manner similar to sandblasting.

Caverns in sea cliffs formed by the ceaseless pounding of the surf or by groundwater flowing through an undersea limestone formation that was hollowed out as the water emptied into the ocean. Sea arches, such as Needle's Eye on Gibraltar Island in western Lake Erie, were created by wave action on limestone promontories with zones of differential hardness. Sometimes sea arches formed by the pounding of the surf as it punched holes into weak chalk beds.

Many arches fashioned by the lateral erosion of a stream flowing around and eventually through the rock evolved into natural bridges (Fig. 175). They

Figure 172 *Shirock, San Juan County, New Mexico.*

(Photo by W. T. Lee, courtesy of USGS)

are narrow, continuous archways of rock, often spanning a ravine or a valley. Rainbow Bridge near Lake Powell just north of the Utah-Arizona border is the world's largest natural bridge. Over the years, sections have broken off and fallen to the ground, prompting fears that in its weakened state, it might collapse altogether.

Natural bridges are the product of erosion and weathering of resistant rocks, such as sandstone or limestone, combined with layers that resist erosion in varying degrees. If a resistant layer overlies a softer bed, it forms a protective cap. When vertical joints or fractures penetrate the softer rock, water flows through and erodes the rock, undercutting the resistant capping layer.

Many bridges form on narrow ridges. In sandstone formations, natural bridges developed from rock shelter caves where part of the roof collapsed and large blocks of rock broke away. Natural bridges were also created by a slower process of spalling, or chipping away grain by grain, leaving a narrow part of the cave roof intact.

In limestone terrain, natural bridges formed in tunnels excavated by groundwater solutions where the tunnel roof collapsed. The most famous example in the United States is Natural Bridge in west-central Virginia. Sometimes,

Figure 173 *Columnar jointing in a massive basalt flow at Devils Postpile National Monument, Madera County, California.*

(Photo by F. E. Matthes, courtesy of USGS)

Figure 174 *Gothic Arch in Navajo sandstone resting on Kayenta sandstone, Garfield County, Utah.*

(Photo by H. E. Gregory, courtesy of USGS)

a large detached block of rock falls or tilts to bridge the gap between two other rock units. Even petrified tree trunks are known to construct natural bridges (Fig. 176). Other small bridges or arches are associated with sinkholes.

KARST TERRAIN

Limestone and other soluble materials underlie large portions of the world. As groundwater percolates through these formations, it dissolves soluble minerals such as calcite, forming large cavities or caverns. The collapsing strata overlying the caverns puncture the ground with sinkholes up to 300 feet or more wide and 100 feet or more deep. At other times, the land surface settles slowly and irregularly. Although the formation of sinkholes is a natural phenomenon, the withdrawal or disposal of water often accelerates the process.

The pits created by dissolving soluble subterranean materials produces a pockmarked landscape known as karst terrain. The name is derived from the region of Karst on the coast of Slovenia famous for its numerous caves. Karst terrain is generally found in areas with moderate-to-abundant rainfall. Throughout the world, some 15 percent of the land surface rests on karst terrain occupied by millions of sinkholes.

Karst terrain and caverns in the United States are mainly located in the southeastern and midwestern sections of the country. Karst terrain also covers

Figure 175 A natural bridge in Navajo Sandstone, Garfield County, Utah.

(Photo by H. E. Gregory, courtesy of USGS)

some portions of the Northeast and West. In Alabama, where limestone and other soluble sediments overlie nearly half the state, thousands of sinkholes pose serious problems for highways and construction projects (Fig. 177). About a third of Florida is underlain by eroded limestone at shallow depths, where sinkholes commonly exist.

Sometimes the land surface's settling causes extensive damage to structures built over cavities formed by dissolving soluble minerals. A dramatic example of this phenomenon occurred in May 1981. A sinkhole 350 feet wide and 125 feet deep suddenly opened up in Winter Park, Florida, collapsing a portion of the town. On December 12, 1995, heavy rainfall and a sewer pipe break in San Francisco, California created a huge sinkhole as deep as a 10-story building. It swallowed a million-dollar house and threatened dozens more.

Karst plains are flat areas with karst features in regions of nearly horizontal limestone strata. A blind valley is a river valley in karst terrain that ends abruptly where the stream disappears underground, called a swallow hole. During heavy rains, a blind valley might become a temporary lake. A type of blind valley called a karst valley forms by the coalescence of several sinkholes. Sinkholes often fill with water and become small, permanent lakes.

In the shallow seas around the Bahamas southeast of Florida, water-filled sinkholes produce large, dark pools of deep seawater called blue holes. They formed during the Ice Age when the ocean dropped several hundred feet, exposing sections of the ocean floor well above sea level. Acidic rainwater that was seeping into the ground dissolved the limestone bedrock and created vast subterranean caverns. Under the weight of the surface rocks, the roofs of the caverns collapsed, exposing huge, gaping pits. When the glaciers melted at the end of the Ice Age, the area refilled with seawater as the ocean rose to near its current level. Much fear and superstition surrounds blue holes because they often exhibit strong eddy currents or whirlpools that can swamp small boats during incoming and outgoing tides.

Figure 176 *Petrified log bridge, Petrified Forest National Monument, Arizona.*

(Photo by R. B. Marshall, courtesy of USGS)

215

Figure 177 *A sinkhole at a construction site in Jefferson County, Alabama.*

(Courtesy of USGS)

The karst terrain in the jungles of Mexico's Yucatán Peninsula displays a bizarre realm of giant undersea caverns and sinkholes of astonishing beauty. Miles of twisting passages 100 feet beneath the ocean link the caves. Strange, sightless creatures occupying the deepest recesses of the caves are blind after generations have lived in utter darkness. The sinkholes formed when the upper surface of a limestone formation collapsed, exposing the watery world beneath the jungle floor.

The underlying limestone is honeycombed with long tunnels, some several miles long, and huge caverns that could easily hold several houses. Like surface caves, the Yucatán caverns contain a profusion of icicle-shaped formations of stalactites hanging from the ceiling and stalagmites clinging to the floor. The formations also include delicate, hollow stalactites called soda straws that took millions of years to create but are destroyed in mere moments by careless divers exploring the caves.

Off the northern coast of the Yucatán Peninsula is a huge meteorite crater called the Chicxulub Structure, named for a small village at its center that means "the devil's tail" in Mayan. It is the largest known crater on Earth, measuring from 110 to 185 miles wide. It lies beneath about a mile of sediments. The impact structure dates precisely to the end of the dinosaur era about 65 million years ago and is thought to have caused the giant beast's extinction. The buried crater is outlined by an unusual concentration of sinkholes. The impact structure forms a circular fracture system that acts as an underground river. The cavity formation in the sinkholes extends to a depth of about 1,000 feet. Its permeability causes the ring to act as a conduit used for carrying water to the sea.

STREAMBED STRUCTURES

Potholes are commonly exposed on ancient streambeds (Fig. 178) and are seen on modern river bottoms. They are generally smooth-sided circular or elliptical holes in hard bedrock. Many potholes are broader at the bottom than at the top, which often has an undercut rim. The world's largest pothole is near Archbald, Pennsylvania and measures 42 feet wide and nearly 50 feet deep. Huge, rounded boulders lying on the bottom suggest they were responsible

Figure 178 *Potholes in sandstone at Squaw shoals just above the mouth of Blue Creek, Tuscaloosa County, Alabama.*

(Photo by C. Butts, courtesy of USGS)

217

for its creation. Torrents of water from a melting Ice Age glacier whirled the rocks around in the hole, abrading the pit to make it wider and deeper.

Another large pothole measuring nearly 40 feet across formed in the bed of the Deerfield River in Shelburne Falls, Massachusetts. It sits in the midst of several smaller potholes cut into the hard bedrock. Excellent examples of potholes up to 5 feet in diameter and 30 feet deep are found on Moss Island in Little Falls, New York. Many of the best potholes lie below dams, exposing the hole-ridden bedrock where strong rapids once flowed. Occasionally, potholes formed on overhanging ledges cut completely through the rock, leaving steeply inclined tunnels.

Potholes described as "glacial" are often located in the northern regions. The glaciers do not actually cut potholes themselves but release vast quantities of water when they melt. In addition, river channels overflowing with meltwater cause severe erosion and potholes. Another indirect effect glaciers have on pothole formation occurs during the drainage of large lakes fed by glacial meltwater. The gradients of streams flowing into the lakes greatly steepen, during which time they erode downward-cutting potholes.

Deep in the Altai Mountains of southern Siberia, perhaps the greatest flood ever to wash over the Earth was unleased during the melting of the great ice sheets around 14,000 years ago, near the end of the last ice age. A glacier cutting across the Chuja Valley created a thick ice dam. This dam held back a large lake nearly 3,000 feet deep that contained some 200 cubic miles of water. When the ice dam broke and the lake burst through, a tremendous deluge of glacial meltwater rushed into the narrow river valley possibly over a period of several days.

The water, reaching as much as 1,500 feet at its height, raced along the Chuja River Valley at 90 miles per hour. The massive flooding formed oddly ripped terrain in the valley and nearby regions. It also left gigantic gravel bars and huge ripples in the gravel beds on the valley floor and surrounding areas. The oddly rippled terrain is similar to the ripple marks seen in river bottoms, only the scale is magnified tremendously. Up to tens of yards occur between crests.

On the border between Idaho and Montana, a gigantic ice dam held back an enormous body of water called Lake Missoula. This lake was hundreds of miles wide, as much as 1,800 feet deep, and contained up to 600 cubic miles of water. Between 15,000 and 13,000 years ago, the dam repeatedly burst, sending massive floods of glacial meltwater gushing toward the Pacific Ocean. This flooding was hundreds of times greater than any flood on the Mississippi River. Along the way, the floodwaters carved out one of the most peculiar landscapes on Earth, known as the Channeled Scablands (Fig. 179). Their distinctive suite of landforms found nowhere else attests to the immense stream power of glacial floods.

The outburst floods occurred when the level of the water below the glacier reached nine-tenths the height of the dam. At that point, the sub-

Figure 179 *The Grand Coulee Dam on the Columbia River in north-west Washington, showing tortured Scablands terrain.*

(Courtesy of USGS)

merged edge of the dam became slightly buoyant and rose, allowing water to leak out underneath. In a few weeks, the running water melted a progressively larger tunnel through the ice until a trickle became a torrent. A huge boil of water was forced out under high pressure, producing a giant, gushing fountain at the snout of the glacier.

When the tunnel through the ice dam grew significantly large, its roof collapsed and created an open channel through which the lake completely drained in a matter of only a few days. Eventually, the southward–creeping ice lobe stemmed the hemorrhage of water and initiated the process all over again. The Missoula floods scarred the face of the land, carving huge canyons and potholes into thick lava formations, resulting in scabland typography. Their presence suggests that catastrophic flooding might account for similar landforms in other parts of the world.

The type section for this unusual topography is the Columbia Basalt Plateau of eastern Washington. The area is broken into a maze of mesas, buttes, and canyons formed by the powerful erosive action of rapidly flowing glacial meltwaters. The most striking features of this region include reticulated (net-like) stream channels, abandoned canyons, hanging valleys, and permanently dry waterfalls.

THE STRANGEST VOLCANOES

In the Cascade Mountains of Oregon is the site of a gigantic volcanic gas explosion that excavated a huge crater named Hole in the Ground. It is a perfectly circular pit several thousand feet across with a rim raised a few hundred feet above the surrounding terrain. Mysteriously, most of the crater lacks vegetation possibly due to poisonous gases that spilled out during the eruption. Another similar structure called Ubehebe Crater is one of the most impressive sights in Death Valley. It exploded into existence about 1,000 years ago when molten basalt came into contact with the shallow groundwater table and flashed into steam.

Perhaps the strangest volcanic eruption occurred under a glacier in Iceland in 1918, unleashing a massive flood of meltwater called a jokulhlaup that was known to Icelanders since the 12th century. In a matter of days, the underglacier eruption released up to 20 times more water than the flow of the Amazon, the world's largest river. A similar underice eruption on September 30, 1996 melted through the 1,700-foot-thick ice cap and sent massive floodwaters and icebergs dashing to the sea a month later. For a short while, this formed the second largest river in the world. It destroyed three bridges, telephone lines, and the only highway running along Iceland's southern coast, with damages estimated at $15 million.

A glacier burst is a sudden release of meltwater from a glacier or subglacial lake. Water accumulates in depressions within the ice margins and erupts through the ice barrier, sometimes resulting in a catastrophic flood. The process of accumulation and release might occur at almost regular intervals. The phenomenon occurs most commonly in Iceland, where it accompanies volcanic or fumarolic (volcanic steam) activity.

Water gushing from an underglacier eruption carves out an enormous ice cave. Geothermal heat beneath the ice creates a large reservoir of meltwater as much as 1,000 feet deep. A ridge of rock acts as a dam to hold back the water. The sudden breakage of the dam causes the flow of water to form a long channel under the ice. Outwash streams of meltwater flowing from a glacier also carve ice caves that can be followed far upstream.

The gigantic ice sheets atop Antarctica hide many volcanoes. Underice eruptions as much as a mile or more beneath the glacier can produce massive floods of meltwater. Antarctica is noted for its unusual volcanoes. The most active is Mount Erebus (Fig. 180), a smoldering mountain that rises 12,500 feet above Ross Island. Its depths were explored in early 1993 by the eight-legged, spiderlike Dante robot, which took a perilous journey down the volcanic vent to collect rock and gas samples.

Several volcanoes puncture the ice of West Antarctica. Many of Antarctica's dormant volcanoes are buried within the ice, and extensive volcanic deposits underlie the ice sheets. When active volcanoes erupt beneath the ice, they spout great floods of meltwater. This mixes with the underlying sediment, forming glacial till tens of feet thick. Basalt erupted beneath the glacial ice produces volcanic rocks called hyaloclastics. They are pillow lavas and pillow breccias, which are unique, quickly frozen forms of lava.

Figure 180 *Ice and snow formations at the foot of Mount Erebus, Antarctica.*

(Photo by W. A. Davis, courtesy of U.S. Navy)

While flying over the Antarctic Ross Ice Shelf in February 1993, scientists noticed a round depression in the ice roughly 4 miles wide and 160 feet deep. Only an active volcano erupting under the glacier could have melted such a large area of ice. By using radar to penetrate the ice sheet, a 4-mile-wide, 2,100-foot-high volcano was discovered under more than a mile of ice. The volcano sits in the middle of a giant caldera 14 miles wide within a rift valley, where the Earth's crust is being stretched apart and hot rock from the mantle is rising to the surface.

Although this was the first active volcano found under the Antarctic ice, it is unlikely the only one. Satellite images have revealed other circular depressions in the ice, suggesting many more volcanoes are lurking beneath the glaciers. The volcanoes could produce enough heat to melt the base of the ice sheet. This would allow ice streams tens of miles wide to flood into the sea, setting the stage for the collapse of the West Antarctic ice sheet. The surge of ice into the ocean would raise global sea levels, and the resulting inundation would radically alter the landscape of the world.

CONCLUSION

One final word needs to be said about future landforms. The Earth changes constantly. As seen in chapter 1, it was much different in the past than it is today. The Earth will continue to change and be even more different in the future. In just a few million years from now, major alterations in the Earth's surface will take place. For example, California westward of the San Andreas Fault will continue to trek northward, finally ending just below Alaska, as the plate upon which it rides plunges down the Aleutian Trench.

At the same time, Baja, California, which rifted apart from mainland Mexico 6 million years ago, will scoot along the west coast of the United States, finally coming to rest just below present-day Canada. Other scraps of land will also continue to accrete to the North American continent, continually enlarging the landmass.

In other parts of the world, the East African Rift Valley will eventually widen and flood with seawater to form a new subcontinent similar to Madagascar, which broke away from Africa about 125 million years ago and became an isolated landmass. The newly formed crustal fragment will then drift toward India, possibly colliding with the subcontinent. India itself will continue to press against South Asia, distorting the Asian continent and causing tremendous earthquakes to rumble across the countryside. The crumpling of the crust will continue to raise the Himalayas and Tibetan Plateau. Huge rivers draining the region will add new land to the Gulf of Bengal, which will continue to be plagued with powerful typhoons. Africa and Arabia will continue diverging from each other, ever widening the Red Sea and the Gulf of Aden and separating Saudi Arabia from Africa.

The Atlantic Ocean will continue to widen at the expense of the Pacific, as new ocean floor is added at the Mid-Atlantic Ridge by seafloor spreading. The Pacific plate will continue to shrink as the ocean floor gets swallowed up by subduction zones in the western Pacific. This jostling of crustal plates will shift the positions of the continents (Fig. 181) The Hawaiian Islands in the central Pacific will be provided with their newest volcanic island, named Loihi. It will rise some 11,000 feet from the ocean floor and emerge in a huge fiery outburst. The rest of the island chain, however, will disappear beneath the ocean waves. Volcanoes throughout the rest of the world will erupt one after another as the Earth vents its excess heat to the surface during a time of rapid continental movements.

South America will separate from North America and drift into the South Pacific. The opening of the Panama Isthmus will allow westward-flowing currents from the Atlantic into the Pacific, shutting off the warm Gulf Stream circum-Atlantic current. This will place Europe into a deep freeze. The loss of this great heat-transfer mechanism will initiate an unprecedented ice age. Its massive glaciers will spread out from the polar regions and plow up real estate far southward, burying cities like St. Louis and London under a sheet of ice at least one-mile thick. The drop in sea level several hundred feet will advance the shoreline in places hundreds of miles seaward, altering the shapes of the continents (Fig. 182).

Figure 181 *The position of the continents 50 million years from now. Darkened areas reveal present positions.*

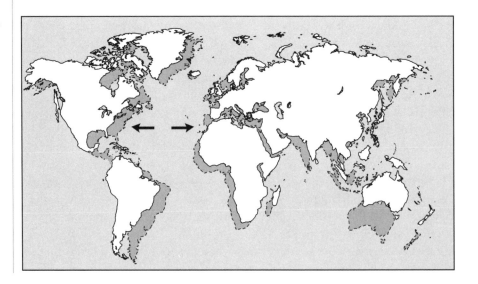

Figure 182 *Extended shoreline during the height of the Ice Age.*

When the glaciers finally retreat, they will leave behind a huge desert wasteland comprised of long, sinuous eskers, thousands of hillocks of glacial till called drumlins, and expansive glacial lakes that dwarf even the present Great Lakes. The outwash of glacial meltwater gorged with sedimentary debris will flood over the continents, carving out new river channels and clogging old streams with mounds of sand and gravel. Plants and animals forced to retreat southward to avoid the harsh ice age conditions will slowly return to recolonize land once claimed by the glaciers.

The African and Eurasian Plates will continue to press against each other. The Mediterranean Sea will be caught in the middle and squeezed dry. New mountain ranges will rise up out of thick sediments that have been accumulating in the Mediterranean Basin for millions of years. Australia will drift northward, possibly colliding with Southeast Asia as the plate it rides upon dives into the gaping Java Trench. At the same time, Japan and Taiwan will crash into eastern China, adding more land to the Asian continent. Antarctica will drift away from the South Polar region, where it has remained stationary for millions of years. The thawing of the ice continent will raise global sea levels several hundred feet and flood coastal areas in places hundreds of miles inland (Fig. 183).

North and South America will continue to slide across the Pacific and then reverse direction and head back toward Eurasia as a new subduction zone opens up just outside the continental shelf in the western Atlantic. The Mid-Atlantic Ridge will become an extinct rift zone and will ultimately become consumed by the subduction zone. Eventually, all continents will reunite into a single, large supercontinent called Neopangaea. Then, the process of continental breakup and drift will begin again, forming new continents and oceans that will bear no resemblance to those of today. The Earth will again become a strange new world.

Figure 183 *Areas in Europe that would be flooded if the ice caps melted.*

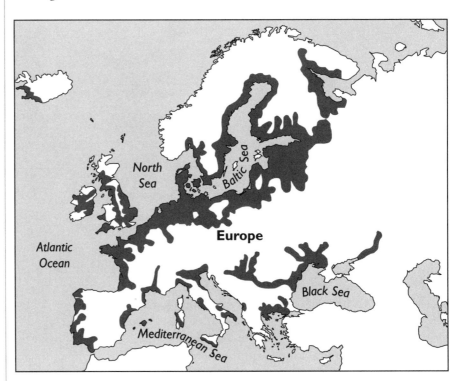

GLOSSARY

aa lava (AH-ah)   a lava that forms large, jagged, irregular blocks

abrasion   erosion by friction, generally caused by rock particles carried by running water, ice and wind

abyss   the deep ocean, generally over a mile in depth

aerosol   a mass of minute solid or liquid particles dispersed in the air

albedo   the amount of sunlight reflected from an object and dependent on its color and texture

alluvium (ah-LUE-vee-um)   stream-deposited sediment

alpine glacier   a mountain glacier or a glacier in a mountain valley

andesite   an intermediate type of volcanic rock between basalt and rhyolite

anticline   folded sediments that slope downward away from a central axis

aquifer (AH-kwe-fer)   a subterranean bed of sediments through which groundwater flows

arches   archlike features in rock formed by erosion

arête (ah-RATE)   a sharp-crested ridge formed by abutting cirques

ash fall   the fallout of small, solid particles from a volcanic eruption cloud

asperite (AS-per-ite)   the point where a fault hangs up and eventually slips, causing earthquakes

asteroid   a rocky or metallic body whose impact on the Earth creates a large meteorite crater

asthenosphere (as-THE-nah-sfir)   a layer of the upper mantle from about 60 to 200 miles below the surface that is more plastic than the rock above and below and that might be in convective motion

astrobleme (as-TRA-bleem)   eroded remains on the Earth's surface of an ancient impact structure produced by a large cosmic body

avalanche (AH-vah-launch)   a slide on a snowbank triggered by vibrations from earthquakes and storms

back-arc basin   a seafloor-spreading system of volcanoes caused by an extension behind an island arc that is above a subduction zone

barrier island   a low, elongated coastal island that parallels the shoreline and protects the beach from storms

basalt (bah-SALT)   a dark, volcanic rock that is usually quite fluid in the molten state

basement rock   subterranean igneous, metamorphic, granitized, or highly deformed rock underlying younger sediments

batholit (BA-the-lith)   the largest of intrusive igneous bodies and more than 40 square miles on its uppermost surface

bedrock   solid layers of rock beneath younger materials

blowout   a hollow caused by wind erosion

blue hole   a water-filled sinkhole

blueschist (BLUE-shist)   metamorphosed rocks of subducted ocean crust exposed on land

breccia (BRE-cha)   a rock composed of angular fragments in a fine-grained matrix

butte   a flat-topped hill with steep slopes

calcite   a mineral composed of calcium carbonate

caldera (kal-DER-eh)   a large, pitlike depression at the summits of some volcanoes and formed by great explosive activity and collapse

calving   formation of icebergs by glaciers breaking off upon entering the ocean

carbonaceous (kar-bah-NAY-shus)   a substance containing carbon, namely sedimentary rocks such as limestone and certain types of meteorites

carbonate   a mineral containing calcium carbonate, such as limestone

carbon cycle   the flow of carbon into the atmosphere and ocean, the conversion to carbonate rock, and the return to the atmosphere by volcanoes

catchment area   the recharge area of a groundwater aquifer

chalk   a soft form of limestone composed chiefly of calcite shells of microorganisms

chert   an extremely hard, fine-grained quartz mineral

circum-Pacific   active seismic regions around the rim of the belt Pacific Plate coinciding with the Ring of Fire

cirque (serk)   a glacial erosional feature, producing an amphitheater-like head of a glacial valley

col (call)   a saddle-shaped mountain pass formed by two opposing cirques

comet a celestial body believed to originate from a cloud of comets that surrounds the sun and develops a long tail of gas and dust particles when traveling near the inner solar system

conglomerate (kon-GLOM-er-ate) a sedimentary rock composed of welded, fine-grained, and coarse-grained rock fragments

continent a landmass composed of light, granitic rock that rides on the denser rocks of the upper mantle

continental drift the concept that the continents have been drifting across the surface of the Earth throughout geologic time

continental glacier an ice sheet covering a portion of a glacier

continental margin the area between the shoreline and the abyss and that represents the true edge of a continent

continental shelf the offshore area of a continent in a shallow sea

continental shield ancient crustal rocks upon which the continents grew

continental slope the transition from the continental shelf to the deep-sea basin

convection a circular, vertical flow of a fluid medium by heating from below; as materials are heated, they become less dense and rise, cool, become more dense, and sink

convergent plate the boundary between crustal plates where the plates come together; generally corresponds to the deep-sea trenches where old crust is destroyed in subduction zones

coral any of a large group of shallow-water, bottom-dwelling marine invertebrates that build reef colonies in warm waters

cordillera (kor-dil-ER-ah) a range of mountains that includes the Rockies, Cascades, and Sierra Nevada in North America and the Andes in South America

core the central part of the Earth, consisting of a heavy iron-nickel alloy; also, a cylindrical rock sample drilled through the crust

correlation (KOR-eh-LAY-shen) the tracing of equivalent rock exposures over a distance, usually with the aid of fossils

craton (CRAY-ton) the stable interior of a continent, usually composed of the oldest rocks

creep the slow flowage of earth materials

crevasse (kri-VAS) a deep fissure in the crust of a glacier

crust the outer layers of a planet's or a moon's rocks

crustal plate a segment of the lithosphere involved in the interaction of other plates in tectonic activity

delta a wedge-shaped layer of sediments deposited at the mouth of a river

desertification (di-zer-te-fa-KA-shen) the process of becoming arid land

desiccated basin (de-si-KAY-ted) a basin formed when an ancient sea evaporated

diapir (DIE-ah-per)   the buoyant rise of a molten rock through heavier rock

dike   a tabular, intrusive body that cuts across older strata

divergent plate   the boundary between crustal plates where the plates move apart, it generally corresponds to the midocean ridges where new crust is formed by the solidification of liquid rock rising from below

dolomite (DOE-leh-mite)   a sedimentary rock formed by the replacement of calcium with magnesium in limestone

domepit   a vertical shaft connecting different levels in a cave

dropstone   a boulder embedded in an iceberg and dropped to the seabed upon melting

drumlin   a hill of glacial debris facing in the direction of glacial movement

dune   a ridge of windblown sediments usually in motion

earth flow   the downslope movement of soil and rock

earthquake   the sudden rupture of rocks along active faults in response to geologic forces within the Earth

East Pacific Rise   a midocean spreading center that runs north-south along the eastern side the Pacific, the predominant location upon which the hot springs and black smokers have been discovered

eolian (ee-OH-lee-an)   a deposit of windblown sediment

erosion   the wearing away of surface materials by natural agents such as wind and water

erratic boulder   a glacially deposited boulder far from its source

escarpment (es-KARP-ment)   a mountain wall produced by the elevation of a block of land

esker (ES-ker)   a curved ridge of glacially deposited material

evaporite (ee-VA-per-ite)   the deposition of salt, anhydrite, and gypsum from evaporation in an enclosed basin of stranded seawater

exfoliation (eks-foe-lee-A-shen)   the weathering of rock, causing the outer layers to flake off

extrusive (ik-STRU-siv)   an igneous volcanic rock ejected onto the surface of a planet or moon

fault   a break in crustal rocks caused by Earth movements

fissure   a large crack in the crust through which magma might escape from a volcano

fjord (fee-ORD)   a long, narrow, steep-sided inlet of a mountainous, glaciated coast

floodplain   the land adjacent to a river that floods during river overflows

fluvial (FLUE-vee-al)   pertaining to being deposited by a river

formation   a combination of rock units that can be traced over a distance

fossil   any remains, impressions, or traces in rock of a plant or animal of a previous geologic age

frost heaving   the lifting of rocks to the surface by the expansion of freezing water

frost polygons   polygonal patterns of rocks from repeated freezing

fumarole (FUME-ah-role)   a vent through which steam or other hot gases escape from underground such as a geyser

gabbro (GA-broe)   a dark, coarse-grained, intrusive igneous rock

geomorphology (JEE-eh-more-FAH-leh-jee)   the study of surface features of the Earth

geothermal   the generation of hot water or steam by hot rocks in the Earth's interior

geyser (GUY-sir)   a spring that ejects intermittent jets of steam and hot water

glacier   a thick mass of moving ice occurring where winter snowfall exceeds summer melting

glacier burst   a flood caused by an underglacier volcanic eruption

glacière (GLAY-sher-ee)   an underground ice formation

Glossopteris (GLOS-op-ter-is)   late Paleozoic tongue-fern living on Gondwana

gneiss (nise)   a banded, coarse-grained, metamorphic rock with alternating layers of different minerals, consisting of essentially the same components as granite

Gondwana (gone-DWAN-ah)   a southern supercontinent of Paleozoic time, comprising Africa, South America, India, Australia, and Antarctica; it broke up into the present continents during the Mesozoic era

graben (GRA-bin)   a valley formed by a down-dropped fault block

granite   a coarse-grained, silica-rich rock, consisting primarily of quartz and feldspars; the principal constituent of the continents, believed to be derived from a molten state beneath the Earth's surface

granulite (GRAN-yeh-lite)   a metamorphic rock comprising continental interiors

greenstone   a green metamorphosed igneous rock of Archean age

groundwater   water derived from the atmosphere and that percolates and circulates below the surface

gypsum (JIP-sem)   a calcium sulfate mineral formed during the evaporation of brine pools

haboob (hey-BUBE)   a violent dust storm or sandstorm

halite   an evaporite deposit composed of common salt

hanging valley   a glaciated valley above the main glaciated valley often forming a waterfall

hiatus (hie-AY-tes)   a break in geologic time due to a period of erosion or nondeposition of sedimentary rock

horn   a peak on a mountain formed by glacial erosion

horst   an elongated, uplifted block of crust bounded by faults

hot spot   a volcanic center with no relation to a plate boundary; an anomalous magma generation site in the mantle

hyaloclastic (hi-AH-leh-KLAS-tic)   basalt lava erupted beneath a glacier

hydrocarbon   a molecule consisting of carbon chains with attached hydrogen atoms

hydrologic cycle   the flow of water from the ocean to the land and back to the sea

hydrology   the study of water flow over the Earth

hydrothermal   relating to the movement of hot water through the crust; also a mineral ore deposit emplaced by hot groundwater

Iapetus Sea (EYE-ap-i-tus)   a former sea that occupied a similar area as the present Atlantic Ocean prior to Pangaea

ice age   a period of time when large areas of the Earth were covered by massive glaciers

iceberg   a portion of a glacier calved off upon entering the sea

ice cap   a polar cover of snow and ice

igneous rocks   all rocks solidified from a molten state

impact   the point on the surface upon which a celestial object has landed, creating a crater

inselberg (IN-sell-berg)   isolated upland standing above the general level of the surrounding plains

interglacial   a warming period between glacial periods

intertidal zone   the shore area between low and high tides

intrusive   any igneous body that has solidified in place below the Earth's surface

island arc   volcanoes landward of a subduction zone, parallel to a trench, and above the melting zone of a subducting plate

isostasy (eye-SOS-teh-see)   a geologic principle that states that the Earth's crust is buoyant and rises and sinks depending on its density

jointing   the production of parallel fractures in rock formations

kame (came)   a steep-sided mound of moraine deposited at the margin of a melting glacier

karst   a terrain comprised of numerous sinkholes in limestone

kettle   a depression in the ground caused by a buried block of glacial ice

kimberlite (KIM-ber-lite)   a volcanic rock composed mostly of peridotite, originating deep within the mantle that brings diamonds to the surface

knoll (nole)   a small, rounded hill

laccolith (LA-keh-lith)   a dome-shaped intrusive magma body that arches the overlying sediments and some times forms mountains

lahar (LAH-har)   a mudflow of volcanic material on the flanks of a volcano

landform   a surface feature of the Earth

landslide   a rapid downhill movement of earth materials triggered by earthquakes and severe weather

lapilli (leh-PI-lie)   small, solid, pyroclastic fragments

lateral moraine   the material deposited by a glacier along its sides

Laurasia (lure-AY-zha) a northern supercontinent of Paleozoic time consisting of North America, Europe, and Asia

lava molten magma that flows out onto the surface

limestone a sedimentary rock consisting mostly of calcite from shells of marine invertebrates

liquefaction (li-kwe-FAK-shen) the loss of support from sediments that liquefy during an earthquake

lithosphere the rocky outer layer of the mantle that includes the terrestrial and oceanic crusts; it circulates between the Earth's surface and mantle by convection currents

lithospheric a segment of the lithosphere involved in the plate interaction of other plates during tectonic activity

loess (LOW-es) a thick deposit of airborne dust

magma a molten rock material generated within the Earth and the constituent of igneous rocks

magnetic field a reversal of the north-south polarity of the magnetic poles

mantle the part of a planet below the crust and above the core, composed of dense rocks that might be in convective flow

mass wasting the downslope movement of rock under the direct influence of gravity

megalithic large stones arranged for various purposes including cultural monuments

mesa an isolated, relatively flat-topped elevated feature larger than a butte and smaller than a plateau

metamorphism (me-teh-MORE-fi-zem) recrystallization of previous igneous, metamorphic, or sedimentary rocks created under conditions of intense temperatures and pressures without melting

meteorite a metallic or stony celestial body that enters the Earth's atmosphere and impacts on the surface

Mid-Atlantic Ridge the seafloor-spreading ridge that marks the extensional edge of the North and South American plates to the west and the Eurasian and African plates to the east

midocean ridge a submarine ridge along a divergent plate boundary where a new ocean floor is created by the upwelling of mantle material

mima mounds piles of sediment caused by earthquakes

monadnock (mah-NAD-nock) an isolated mountain or hill rising above a lowland

moraine (mah-RANE) a ridge of erosional debris deposited by the melting margin of a glacier

moulin (mue-LIN) a cylindrical shaft extending down into a glacier and produced by meltwater

mountain roots the deeper crustal layers under mountains

mudflow the flowage of sediment-laden water

nuée ardente (NU-ee ARE-dent) a volcanic pyroclastic eruption of hot ash and gas

ophiolite (OH-fi-ah-lite) masses of oceanic crust thrust onto the continents by plate collisions

orogen (ORE-ah-gin) an eroded root of an ancient mountain range

orogeny (oh-RAH-ja-nee) a process of mountain building by tectonic activity

outgassing the loss of gas from within a planet as opposed to degassing, the loss of gas from meteorites

overthrust a thrust fault in which one segment of crust overrides another segment for a great distance

oxbow lake a cutoff section of a river meander that forms a lake

pahoehoe lava (pah-HOE-ay-hoe-ay) a lava that forms ropelike structures when cooled

paleontology (pay-lee-on-TAH-logy) the study of ancient life-forms based on the fossil record of plants and animals

Pangaea (pan-GEE-a) an ancient paleozoic supercontinent that included all the lands of the Earth

Panthalassa (pan-THE-lass-a) a great global ocean that surrounded Pangaea

pediment an inclined erosional surface that slopes away from a mountain front and is often covered with alluvium

pegmatite a granite with extremely large quartz and feldspar crystals

peneplain a land surface of slight relief shaped by erosion

peridotite (pah-RI-deh-tite) the most common rock type in the mantle

periglacial referring to geologic processes at work adjacent to a glacier

permafrost permanently frozen ground in the Arctic regions

pillow lava lava extruded on the ocean floor and giving rise to tabular shapes

placer (PLAY-ser) a deposit of rocks left behind by a melting glacier; any ore deposit enriched by stream action

plateau (pla-TOW) an extensive region with a relatively level surface rising abruptly above the adjacent land

plate tectonics the theory that accounts for the major features of the Earth's surface in terms of the interaction of lithospheric plates

playa (PLY-ah) a flat, dry, barren plain at the bottom of a desert basin

pluton (PLUE-ton) an underground body of igneous rock younger than the rocks that surround it and formed where molten rock oozes into a space between older rocks

pothole deep depressions in the bedrock of a fast-flowing stream or beneath a waterfall

pumice volcanic ejecta that is full of gas cavities and extremely light in weight

pyroclastic (PIE-row-KLAS-tik)   the fragmental ejecta released explosively from a volcanic vent

quartzite   a metamorphosed sandstone

radiolarian   microorganisms with shells of silica that comprise a large component of siliceous sediments

radiometric dating   determining the age of an object by chemically analyzing its stable and unstable radioactive elements

recessional morraine   a glacial moraine deposited by a retreating moraine glacier

redbed   red-colored sedimentary rocks indicative of a terrestrial deposition

reef   the biological community that lives at the edge of an island or continent; the shells from dead organisms form a limestone deposit

regression   a fall in sea level, exposing continental shelves to erosion

resurgent caldera   a large caldera that experiences renewed volcanic activity that domes up the caldera floor

rhyolite (RYE-oh-lite)   a volcanic rock that is highly viscous in the molten state and usually ejected explosively as pyroclastics

rhythmite (RITH-mite)   regularly banded deposits formed by cyclic sedimentation

rift valley   the center of an extensional spreading, where continental or oceanic plate separation occurs

rill (ril)   a trench formed by a collapsed lava tunnel

riverine (RI-vah-rene)   relating to a river

roche moutonnée (ROSH mue-tin-ay)   a knobby, glaciated bedrock surface

saltation   the movement of sand grains by wind or water

salt dome   an upwelling plug of salt that arches surface sediments and often serves as an oil trap

sand boil   a geyserlike fountain of sediment-laden water produced by the liquefaction process during an earthquake

sandstone   a sedimentary rock consisting of cemented sand grains

scarp   a steep slope formed by Earth movements

schist (shist)   a finely, layered metamorphic rock that tends to split readily into thin flakes

seafloor spreading   a theory that the ocean floor is created by the separation of lithospheric plates along midocean ridges with new oceanic crust formed from mantle material that rises from the mantle to fill the rift

seamount   a submarine volcano

sedimentary rock   a rock composed of fragments cemented together

seiche (seech)   a wave oscillation on the surface of a lake or landlocked sea

seismic sea wave   an ocean wave generated by an undersea earthquake or volcano; also called tsunami

shield   areas of exposed Precambrian nucleus of a continent

shield volcano   a broad, low-lying volcanic cone built up by lava flows of low viscosity

silica   silicon dioxide, SiO_2, the dominant mineral substance in igneous, metamorphic, and sedimentary rocks

siliceous   refering to a rock that contains abundant silica

sill   an intrusive magma body parallel to planes of weakness in the overlying rock

sinkhole   a large pit formed by the collapse of surface materials undercut by the solution of subterranean limestone

solifluction (SOE-leh-flek-shen)   the failure of earth materials in tundra

stalactite (stah-LAK-tite)   a conical calcite deposit hanging from a cave ceiling

stalagmite (stah-LAG-mite)   a conical calcite deposit growing from a cave floor

strata   layered rock formations; also called beds

striae (STRY-aye)   scratches on bedrock made by rocks embedded in a moving glacier

subduction zone   a region where an oceanic plate dives below a continental plate into the mantle; ocean trenches are the surface expression of a subduction zone

subsidence   the compaction of sediments due to the removal of fluids

surge glacier   a continental glacier that heads toward the sea at a high rate of advance

syncline (SIN-kline)   a fold in which the beds slope inward toward a common axis

taiga (TIE-gah)   an extensive pine forest adjacent to the tundra

talus cone   a steep-sided pile of rock fragments at the foot of a cliff

tarn   a small lake formed in a cirque

tectonic activity   the formation of the Earth's crust by large-scale movements throughout geologic time

tephra (TEH-fra)   all-clastic material from dust particles to large chunks, expelled from volcanoes during eruptions

terrace   a narrow, level plain with a steep front bordering a river

terrane (teh-RAIN)   a unique crustal segment attached to a landmass

Tethys Sea (TEH-this)   the hypothetical midlatitude region of the oceans of Paleozoic and Mesozioc time separating the northern and southern continents of Laurasia and Gondwana

till   sedimentary material deposited by a glacier

tillite   a sedimentary deposit composed of glacial till

transgression   a rise in sea level that causes flooding of the shallow edges of continental margins

trapps (traps)   a series of massive lava flows that resembles a staircase

trench   a depression on the ocean floor caused by plate subduction

tropical glacier   a glacier found on top of a mountain in the tropics

tsunami (sue-NAH-me)   a seismic sea wave produced by an undersea or near-shore earthquake or volcanic eruption

tuff   a rock formed of pyroclastic fragments

tundra   permanently frozen ground at high latitudes and elevations

unconformity   an erosional surface separating younger rock strata from older rocks

uplift   the lifting upward of strata above the prevailing terrain

upwelling   the process of rising as in magma or an ocean current

varves   thinly laminated lake bed sediments deposited by glacial meltwater

ventifact (VEN-teh-fakt)   a stone shaped by the action of windblown sand

volatile   a substance in magma, such as water and carbon dioxide, that controls the type of volcanic eruption

volcano   a fissure or vent in the crust through which molten rock rises to the surface to form a mountain

BIBLIOGRAPHY

CONTINENTAL FORMATION

Dalziel, Ian W. D. "Earth Before Pangaea," *Scientific American* 272 (January 1995): 58–63.

Dickinson, William R. "Making Composite Continents," *Nature* 364 (July 22, 1993): 284–285.

Galer, Stephen J. G. "Oldest Rocks in Europe," *Nature* 370 (August 18, 1994): 505–506.

Hoffman, Paul F. "Oldest Terrestrial Landscape," *Nature* 375 (June 15, 1995): 537–538.

Kunzig, Robert. "Birth of a Nation," *Discover* 11 (February 1990): 26–27.

Moores, Eldridge. "The Story of Earth," *Earth* 6 (December 1996): 30–33.

Shurkin, Joel and Tom Yulsman. "Assembling Asia," *Earth* 4 (June 1995): 52–59

Stager, Curt. "Africa's Great Rift," National Geographic 177 (May 1990): 10–41.

Taylor, S. Ross and Scott M. McLennan. "The Evolution of Continental Crust," *Scientific American* 274 (January 1996): 76–81.

Taylor, Stuart Ross. "Young Earth Like Venus?" *Nature* 350 (April 4, 1991): 376–377.

Weiss, Peter. "Land Before Time," *Earth* 8 (February 1998): 29–33.

TECTONIC PROCESSES

Coffin, Millard F. and Olav Eldholm. "Large Igneous Provinces," *Scientific American* 269 (October 1993): 42–49.

Finkbeiner, Ann. "Terra Infirma," *Discover* 12 (November 1991): 18.

Fischman, Joshua. "Falling Into the Gap," *Discover* 13 (October 1992): 57–63.

Glanz, James. "Erosion Study Finds High Price for Forgotten Menace," *Science* 267 (February 24, 1995): 1088.

Gordon, Richard G. and Seth Stein. "Global Tectonics and Space Geodesy," *Science* 256 (April 17, 1992): 333–341.

Hoffman, Paul F. "Oldest Terrestrial Landscape," *Nature* 375 (June 15, 1995): 537.

Hopkins, Ralph Lee. "Land Torn Apart," *Earth* 6 (February 1997): 37–41.

Johnston, Arch C. and Lisa R. Kanter. "Earthquakes in Stable Continental Crust," *Scientific American* 262 (March 1990): 68–75.

Monastersky, Richard. "When Mountains Fall," *Science News* 142 (August 29, 1992): 136–138.

Shaefer, Stephen J. and Stanley N. Williams. "Landslide Hazards," *Geotimes* 36 (May 1991): 20–22.

Stein, Ross S. and Robert S. Yeats. "Hidden Earthquakes," *Scientific American* 260 (June 1989): 48–57.

White, Robert S. and Dan P. McKenzie. "Volcanism at Rifts," *Scientific American* 261 (July 1989): 62–71.

Zimmer, Carl. "Landslide Victory," *Discover* 12 (February 1991): 66–69.

MOUNTAIN RANGES

Bird, Peter. "Formation of the Rocky Mountains, Western United States: A Continuum Computer Model," *Science* 239 (March 25, 1988): 1501–1507.

Harrison, T. Mark, et al. "Raising Tibet," *Science* 255 (March 27, 1992): 1663–1670.

Howard, Ken. "Cutting Off the World's Roof," *Earth* 6 (December 1997): 20–21.

Kerr, Richard A. "Urals Yield Secret of a Lasting Bond," Science 274 (October 11, 1996): 181.

———. "Did Deeper Forces Act To Uplift the Andes?" *Science* 269 (September 1, 1995): 1215–1216.

———. "Making Mountains with Lithospheric Drips," *Science* 239 (February 26, 1988): 978–979.

Larson, Roger L. "The Mid-Cretaceous Superplume Episode," *Scientific American* 272 (February 1995): 82–86.

Monastersky, Richard. "Mountains Frozen in Time," *Science News* 148 (December 23, 30, 1995): 431.

————. "What's Holding Up the High Sierras?" *Science News* 138 (December 15, 1990): 380.

Murphy, J. Brendan and R. Damian Nance. "Mountain Belts and the Supercontinent Cycle," *Scientific American* 226 (April 1992): 84–91.

Pinter, Nicholas and Mark T. Brandon. "How Erosion Builds Mountains," *Scientific American* 276 (April 1997): 74–79.

Ruddiman, William F. and John E. Kutzbach. "Plateau Uplift and Climate Change," *Scientific American* 264 (March 1991): 66–74.

Svitil, Kathy A. "The Coming Himalayan Catastrophe," *Discover* 16 (July 1995): 81–85.

VOLCANIC TERRAIN

Berreby, David. "Barry Versus the Volcano," *Discover* 12 (June 1991): 61–67.

Dvorak, John J., Carl Johnson, and Robert I. Tilling. "Dynamics of Kilauea Volcano," *Scientific American* 267 (August 1992): 46–53.

Hon, Ken and John Pallister. "Wrestling with Restless Calderas and Fighting Floods of Lava," *Nature* 376 (August 17, 1995): 554–555.

Kerr, Richard A. "Volcanoes With Bad Hearts Are Tumbling Down All Over," *Science* 264 (April 29, 1994): 660.

Lewis, G. Brad. "Island of Fire," *Earth* 4 (October 1995): 32–33.

Mojlenbrock, Robert H. "Mount St. Helens, Washington," *Natural History* (June 1990): 27–28.

Monastersky, Richard. "When Kilauea Crumbles," *Science News* 147 (April 8, 1995): 216–218.

Mouginis-Mark, Peter J. "Volcanic Hazards Revealed by Radar Interferometry," *Geotimes* 39 (July 1994): 11–13.

Oliwenstein, Lori. "Lava and Ice," *Discover* 13 (October 1992): 18.

Pendick, Daniel. "Ashes, Ashes, All Fall Down," *Earth* 6 (February 1997): 32–33.

Rampino, Michael R. and Richard B. Stothers. "Flood Basalt Volcanism During the Past 250 Million Years," *Science* 241 (August 5, 1988): 663–667.

Stimac, Jim. "The Strangest Place on Earth," *Earth* 6 (April 1998): 72–75.

Wickelgren, Ingrid. "Simmering Planet," *Discover* 11 (July 1990): 73–75.

RIVERINE TOPOGRAPHY

Adler, Jerry. "Troubled Waters," *Newsweek* (June 26, 1993): 21–27.

Birnbraum, Stephen. "The Grand Canyon," *Good Housekeeping* (November 1990): 150–152.

Evans, Diane L. "Earth From Sky," *Scientific American* 271 (December 1994): 70–75.

Glanz, James. "Erosion Study Finds High Price for Forgotten Menace," *Science* 267 (February 24, 1995): 1088.

Hopkins, Ralph Lee. "Land Torn Apart," *Earth* 8 (February 1997): 37–41.

Mack, Walter N. and Elizabeth A. Leistikow. "Sands of the World," *Scientific American* 275 (August 1996): 62–67.

Macilwain, Colin. "Conservationists Fear Defeat on Revised Flood Control Policies," *Nature* 365 (October 7, 1993): 468.

Miller, Martin. "Missing Time," *Earth* 4 (October 1995): 58–60.

Monastersky, Richard. "Rivers in a Greenhouse World," *Science News* 137 (June 9, 1990): 365.

Nichols, Frederic H., et al. "The Modification of an Estuary," *Science* 231 (February 7, 1986): 567–573.

Pestrong, Ray. "It's About Time," *Earth Science* 42 (Summer 1989): 14–15.

COASTAL GEOGRAPHY

Folger, Tim. "Waves of Destruction," *Discover* 15 (May 1994): 68–73.

Fritz, Sandy. "The Living Reef," *Popular Science* 246 (May 1995): 48–51.

Kerr, Richard A. "From One Coral Many Findings Blossom," *Science* 248 (June 15, 1990): 1314.

Lockridge, Patricia A. "Volcanoes and Tsunamis," *Earth Science* 42 (Spring 1989): 24–25.

Macilwain, Colin. "Conservationists Fear Defeat on Revised Flood Control Policies," *Nature* 365 (October 7, 1993): 478.

Maslin, Mark. "Waiting for the Polar Meltdown," *New Scientist* 139 (September 4, 1993): 36–41.

Monastersky, Richard. "Temperatures on the Rise in Deep Atlantic," *Science News* 145 (May 7, 1994): 295.

———. "Seawall's Seal of Approval," *Science News* 134 (December 17, 1988): 398.

Norris, Robert M. "Sea Cliff Erosion: A Major Dilemma," *Geotimes* 35 (November 1990): 16–17.

Pendick, Daniel. "Waves of Destruction," *Earth* 6 (February 1997): 27–29.

Parfit, Michael. "Polar Meltdown," *Discover* 10 (September 1989): 39–47.

Svitil, Kathy A. "The Sound and the Fury," *Discover* 16 (January 1995): 75

DESERT FEATURES

Ford, J. P., et al. "Faults in the Mojave Desert, California, as Revealed on Enhanced Landsat Images," *Science* 248 (May 25, 1990): 1000–1003.

Hecht, Jeff. "Death Valley Rocks Skate on Thin Ice," *New Scientist* 147 (September 30, 1995): 19.

Mack, Walter N. and Elizabeth A. Leistikow. "Sands of the World," *Scientific American* 275 (August 1996): 62–67.

Monastersky, Richard. "Sahara Dust Blows Over United States," *Science News* 148 (December 23, 30, 1995): 431.

Nuhfer, Edward B. "What's a Geologic Hazard?" *Geotimes* 39 (July 1994): 4.

Pennisi, Elizabeth. "Dancing Dust," *Science News* 142 (October 3, 1992): 218–220.

Peterson, Ivars. "Digging Into Sand," *Science News* 136 (July 15, 1989): 40–42.

Raloff, Janet. "Holding on to the Earth," *Science News* 144 (October 1993): 280–281.

Repetto, Robert. "Deforestation in the Tropic," *Science American* 262 (April 1990): 36–42

Stevens, William K. "Threat of Encroaching Deserts May Be More Myth Than Fact," *The New York Times* (January 18, 1994): C1, C10.

Szelc, Gary. "Where Ancient Seas Meet Ancient Sand," *Earth* 5 (December 1997): 78–81.

Zimmer, Carl. "How to Make a Desert," *Discover* 16 (February 1995): 51–56.

ARCTIC GEOLOGY

Alley, Richard B. and Michael L. Bender. "Greenland Ice Cores Frozen in Time," *Scientific American* 279 (February 1998): 80–85.

Clark, Peter U. "Fast Glacier Flow Over Soft Beds," *Science* 267 (January 6, 1995): 43–44.

Fahnestock, Mark. "An Ice Shelf Breakup," *Science* 271 (February 9, 1996): 775–776.

Horgan, John. "Antarctic Meltdown," *Scientific American* 268 (March 1993): 19–28.

Krantz, William B., Kevin J. Gleason, and Nelson Cain. "Patterned Ground," *Scientific American* 259 (December 1988): 68–76.

Leal, Jose H. "Double-decked Ice Shelf," *Sea Frontiers* 39 (January/February 1993): 21.

MacKenzie, Debora. "Did Northern Forests Stave off Global Warming?" *New Scientist* 139 (September 11, 1993): 6.

Monastersky, Richard. "Satellite Radar Keeps Tabs on Glacial Flow," *Science News* 144 (December 4, 1993): 373.

Pearce, Fred. "Does Polluted Air Keep the Arctic Cool?," *New Scientist* 144 (October 29, 1994): 19.

Pollack, Henry N. and David S. Chapman. "Underground Records of Changing Climate," *Scientific American* 268 (June 1993): 44–50.

Stevens, Jane E. "Life On a Melting Continent," *Discover* 16 (August 1995): 71–75.

GLACIAL LANDSCAPES

Allen, Joseph Baneth and Tom Waters. "The Great Northern Ice Sheet," *Earth* 4 (February 1995): 12–13

Appenzeller, Tim. "After the Deluge," *Scientific American* 261 (December 1989): 22–26.

Edwards, Rob. "Northern Exposure," *New Scientist* 147 (September 9, 1995) 32–36.

Hallet, Bernard and Jaakko Putkonen. "Surface Dating of Dynamic Landforms: Young Boulders on Aging Moraines," *Science* 265 (August 12, 1994): 937–940.

Horgan, John. "The Big Thaw," *Scientific American* 274 (November 1995): 18–20.

Kimber, Robert. "A Glacier's Gift," *Audubon* 95 (May/June 1993): 52–53.

Mollenhauer, Erik. "Glacier On the Move," *Earth Science* 41 (Spring 1988): 21–24.

Monastersky, Richard. "Stones Crush Standard Ice History," *Science News* 145 (January 1, 1994): 4.

———. "Hills Point to Catastrophic Ice Age Floods," *Science News* 136 (September 30, 1989): 213.

Moran, Joseph M., Ronald D. Stieglitz, and Donn P. Quigley. "Glacial Geology," *Earth Science* 41 (Winter 1988): 16–18.

Peltier, Richard W. "Ice Age Paleotopography," *Science* 265 (July 8, 1994): 195–201.

Waters, Tom. "A Glacier Was Here," *Earth* 4 (February 1995): 58–60.

UNIQUE LANDFORMS

Aydin, Atilla. "Evolution of Polygonal Fracture Patterns in Lava Flows," *Science* 239 (January 29, 1988): 471–475.

Baker, Victor R., Gerardo Benito, and Alexey N. Rudoy. "Paleohydrology of Late Pleistocene Superflooding, Atlay Mountains, Siberia," *Science* 259 (January 15, 1993): 348–350.

Bolton, David W. "Underground Frontiers," *Earth Science* 40 (Summer 1987): 16–18.

Folger, Tim. "The Biggest Flood," *Discover* 15 (January 1994): 37–38.

Goodwin, Bruce K. "The Hole Truth," *Earth Science* 41 (Summer 1988): 23–25.

Gorman, Christine. "Subterranean Secrets," *Time* (November 30, 1992): 64–67.

Hanson, Michael C. "Ohio Natural Bridges," *Earth Science* 41 (Winter 1988): 10–11.

Lipske, Mike. "Wonder Holes," *International Wildlife* 20 (January/February 1990): 47–51.

Monastersky, Richard. "Volcanoes Under Ice: Recipe for a Flood," *Science News* 150 (November 23, 1996): 327.

Naeye, Robert. "The Strangest Volcano," *Discover* 15 (January 1994): 38.

Palmer, Arthur N. "Paleokarst Yields Diagenetic Clues," *Geotimes* 40 (September 1995): 9.

Pendick, Daniel. "When the Dam Breaks," *Earth* 6 (February 1997): 30–31.

Petrini, Cathy. "Heart of the Mountain," *Earth Science* 41 (Summer 1988): 14–18.

INDEX

Boldface page numbers indicate extensive treatment of a topic. *Italic* page numbers indicate illustrations or captions. Page numbers followed by *m* indicate maps; *t* indicate tables; *g* indicate glossary.